More Praise for *Regulating from Nowhere*

"This study, deeply and thoroughly informed by philosophical, legal, and political analysis, presents the practical and moral wisdom we need to assess the history and to project the future of environmental law."
—Mark Sagoff, University of Maryland

"*Regulating from Nowhere* is a bold, intriguing, innovative, and beautifully written work. It is extremely well thought out and makes an exceedingly important and original contribution to the field. No other work takes on the problems and issues surrounding environmental regulation with such a deep and sophisticated knowledge and understanding of the ideas of philosophy and political theory."
—Amy Sinden, Associate Professor, Temple University Beasley School of Law

"Professor Kysar provides an analytically compelling critique of cost-benefit analysis and a graceful defense of traditional precautionary approaches to protecting health, safety, and the environment."
—Thomas O. McGarity, author of *The Preemption War*

"This book makes an important, creative, and highly original contribution to the ongoing debate between advocates of a utilitarian environmental law based on cost-benefit analysis and those who support the normative judgments at the heart of the environmental statutes passed in the 1970s. In *Regulating from Nowhere*, Kysar adeptly and vigorously reestablishes a place for moral judgment in crafting environmental policy."
—David M. Driesen, University Professor, College of Law, Syracuse University

REGULATING FROM NOWHERE

DOUGLAS A. KYSAR

Regulating from Nowhere

ENVIRONMENTAL LAW AND

THE SEARCH FOR OBJECTIVITY

Yale UNIVERSITY PRESS

NEW HAVEN AND LONDON

Set in Scala type by Westchester Book Group, Danbury, Connecticut.
Printed in the United States of America by Sheridan Books, Ann Arbor, Michigan.

Library of Congress Cataloging-in-Publication Data

Kysar, Douglas A.
  Regulating from nowhere : environmental law and the search for objectivity / Douglas A. Kysar.
       p. cm.
  Includes bibliographical references and index.
  ISBN 978-0-300-12001-1 (pbk. : alk. paper)   1. Environmental law—Philosophy.
2. Environmental law—United States.   I. Title.
  K3585.K97 2010
  344.04'6—dc22

                                                                2009044763

A catalogue record for this book is available from the British Library.

This paper meets the requirements of ANSI/NISO Z39.48–1992 (Permanence of Paper).

10 9 8 7 6 5 4 3 2 1

Our meddling intellect
Mis-shapes the beauteous forms of things:—
We murder to dissect.

—WILLIAM WORDSWORTH *(1770–1850)*

CONTENTS

FOR THE LATE PHILOSOPHER EMMANUEL LEVINAS, the entire Torah could be found, intricately enfolded, within a simple command— "Thou shalt not kill"—which issues without words from the face of an other who first animates our consciousness and invites us into subjectivity. His was another way of saying that we do not know what all we might lose, once we cease to view the doing of harm as a personal and irredeemable failing, even if an unavoidable one. For Levinas, we are born into responsibility we can neither shed nor fulfill.

Echoes / Cronon's Sentiments...

This book was written to oppose a way of thinking that has come to dominate environmental law and that stands in deep tension with these ideas. The book's central argument is that in order to recognize and consider our most pressing environmental questions, we must inhabit an agent-relative conception of governance—one in which our *particular* political community is acknowledged as a reflective and responsible subject, imbued with an identity, a history, and a legacy still in formation, and capable of reasoning through its obligations not only to its own citizens but also to those who reside within other communities. So too must those other communities be conceptualized as more than mere empirical givens or rationalist automata: The other and the environment of the other must again become of paramount and searching concern, whether the other we glimpse is removed geographically, temporally, or biologically.

"environment of the other"

Dominant ways of thinking about environmental law and policy obscure these needs. They aspire to objectivity and optimality, yet they distort and confuse. Decisions about how to behave in the face of uncertainty concerning potentially devastating consequences of human action are masked by statistical procedures that, notwithstanding their technical sophistication, depend on flatly unwarranted assumptions about the physical world. Society's level of commitment to the avoidance of harm is derived from real or contrived market decisions in a process that, at best, is beset by methodological difficulties and, at worst, misconstrues the very nature of the problem at issue. Perplexing and unavoidable choices about how to influence the environmental context of future generations—choices that will, in a nontrivial sense, influence the content and quality of their lives—are elided through an elaborate mathematical fiction, the discount rate.

The discount rate as a mathematical fiction

These limitations are most pronounced in the context of regulatory cost-benefit analysis, a policy-making approach that achieves its appearance of case-specific rationality at the price of insensitivity to context and to longer-term (systemic rationality.)This book argues that policy makers cannot be content with the local equilibriums identified by conventional cost-benefit analyses but instead must aim to alter—over time and in rather dramatic macroscale ways—the economic and technological forces that combine to structure any given policy context with its microscale snapshot that seems to admit of only one "optimal" solution. To effectively serve such ambitious aims, environmental law must form part of the social glue that binds a political community together in pursuit of long-term and uncertain goals. To serve that function, in turn, laws must have continuity with the concepts, values, and discourses expressed by real people. By literally denying the sacredness of life—and indeed the distinctiveness of anything—dominant ways of approaching environmental law fail these tests.

idea is to expand macroscale snapshot?

*

In preparing this manuscript, I have benefited from the comments and suggestions of far too many colleagues, students, and friends to enumerate. I am especially grateful to several individuals who read and thoughtfully critiqued the entire manuscript—Bruce Ackerman, Rick Brooks, Bernie Meyler, Gus Speth, and two anonymous reviewers for

Yale University Press. Along the way I also received crucial feedback and guidance—often at a point when only truly sympathetic listeners could glean insight from my ill-formed views—from Matt Adler, Bill Alford, Sahand Boorboor, Jules Coleman, Park Doing, Bill Eskridge, Dan Farber, Ron Garet, Ryan Goodman, Bob Hockett, Dan Kahan, Rob Kar, Ryan Keller, Mark Kelman, Rebecca Kysar, Don Maier, Daniel Markovits, Jerry Mashaw, Trevor Morrison, Josh Ober, Jeff Rachlinski, Charles Reich, Susan Rose-Ackerman, J. B. Ruhl, Mark Sagoff, Joanne Scott, Ted Seto, Seana Shiffrin, Sandip Tiwari, Tom Ulen, and Steve Yeazell.

I am deeply indebted to Tom McGarity and Ya-Wei Li, my coauthors on related projects that very much informed the development of this one; some phrases or paragraphs from those projects have likely found their way into this text. Paul Beaton, Ya-Wei Li, Ashley McDowell, Michael Page, William Rinner, Stephanie Tang, and Diana Varat provided outstanding research assistance at various stages of the book's development. Research for this book was supported by Yale Law School, Cornell Law School, and the National Nanotechnology Infrastructure Network. I owe special thanks to Mike O'Malley, whose talents and temperament as an editor are exceptional, and to Jack Borrebach, Kate Davis, Thomas Kozachek, and Alex Larson, who also ably contributed to the book's production. Finally, I am grateful beyond words to Christine Kim, who encountered this project late and who breathed new life into it.

Portions of chapter 3 derive from an article coauthored with Tom McGarity, "Did NEPA Drown New Orleans? The Levees, the Blame Game, and the Hazards of Hindsight," 56 *Duke Law Journal* 179 (2006). Chapter 5 draws from the article "Sustainable Development and Private Global Governance," 83 *Texas Law Review* 2109 (2005) and from a chapter coauthored with Ya-Wei Li, "Regulating from Nowhere: Domestic Environmental Law and the Nation-State Subject," which appeared in volume 2 of the series *The Impact of Globalization on the United States* (2008), edited by Beverly Crawford, Michelle Bertho, and Ed Fogarty. Chapter 6 develops material presented in "Discounting . . . on Stilts, 74 *University of Chicago Law Review* 119 (2007). The research underlying chapter 8 was first presented in "Fish Tales," which appeared in the collection *Alternative Approaches to Regulatory Impact Analysis: A Dialogue Between Advocates and Skeptics of Cost-Benefit Analysis* (2009), edited by Winston

Harrington, Lisa Heinzerling, and Richard Morgenstern. Ideas and arguments throughout this book build upon a lecture first delivered at the Florida State University College of Law and later published as "It Might Have Been: Risk, Precaution, and Opportunity Costs," 22 *Journal of Land Use and Environmental Law* 1 (2006).

REGULATING FROM NOWHERE

INTRODUCTION

BY NOW, the story of modern American environmental law has been redacted into a familiar script, one in which the excesses of our early attempts to regulate the human impact on the environment came to be disciplined by the insights of <u>sound science</u> and <u>economic reasoning</u>, warding off in the process alarmism, inefficiency, and government over-reaching. The script is now so well rehearsed and broadly endorsed that it forms a major component of the United States' intellectual export trade: Through multilateral treaty negotiations and within international forums, such as the World Trade Organization and the Codex Alimentarius Commission, the United States consistently urges the view that efforts to protect the environment and human health must be preceded by an empirical justification derived from a scientific risk assessment and tailored to reflect only the level of environmental or human health protection that is acceptable in light of corresponding costs. When regulation is justified, the story goes, it should be devised in a manner that works with the market, rather than against it, deploying taxes, tradable permits, and other economic instruments that afford maximum flexibility to private actors as they respond to the government's environmental policy dictates. As we learned the hard way, United States officials tell their international counterparts, any departure from this approach invites unnecessarily costly compliance behavior, unwarranted health and

Regulation should work w/ the market

safety concerns, disguised economic protection of domestic industry, and a variety of other regulation-induced ills.

However useful this script may once have been, it now actively impedes efforts to understand and improve our environmental performance. Its logic and conclusions have begun to appear so powerful that we have lost sight of a great deal of practical and moral wisdom that remains alive within our early, "excessive" efforts to conserve natural resources, reduce pollution, save species, and enhance human health and safety. The familiar story has caused, in essence, a *forgetting* of environmental law, a loss of appreciation for why the subject of environmental law is, after all, urgent. Soon enough, the language of instrumentalism that animates our talk of tradeoffs, efficiency, and welfare maximization will become so dominant that we will lose facility altogether with these alternative and once-resonant languages. We will forget that we once talked of environmental rights, rather than of optimal risk tradeoffs; of the grave challenges posed by uncertainty regarding potentially disastrous or irreversible consequences of human action, rather than of risk aversion and the option value of delay; of the stewardship obligations we incur on behalf of future generations, rather than of discounted welfare maximization; and of the responsibility we hold to lead international cooperative endeavors to protect the global biosphere, rather than of competitiveness concerns arising from regulatory differentiation within the world economy. In short, we will forget the richly contoured and sometimes convoluted but always essential moral and political landscape that lends meaning to those aspects of our environmental laws that appear nonsensical from the perspective of economic theory.

Many valuable critiques of the dominant script have been written.[1] This book aims to complement them by illuminating the forgotten wisdom of our traditional, but now much-derided, precautionary approach to environmental governance. Numerous shortcomings of the dominant script will be pointed out along the way, but they will be offered in service of the positive lessons to be had from respectful attention to past experience. Within academic circles, a widely invoked but rarely examined mantra asserts that "it takes a theory to beat a theory." What this mantra has been taken to mean in the environmental law and policy context is that if critics of the dominant script are to succeed, they must do more than simply point out the shortcomings of risk assessment,

[margin note: A forgetting of environmental law?]

[margin note: alternative languages?]

[margin note: Precautionary approach]

cost-benefit analysis, and other allied components of the economic re-form agenda. They must also offer something new and constructive them-selves in the way of environmental law reform.

This is a fine view as far as it goes, but it obscures an important point: _We already had a theory._ It was messy, pluralistic, and pragmatic, and it certainly gave rise to particular regulatory goals and approaches that could be improved upon, but it nonetheless expressed a variety of wisdoms that today are deeply underappreciated. Once restored and il-luminated, these wisdoms explain why so many of our current environ-mental problems appear to us as hopelessly intractable, why international action on climate change, for instance, has been held up for three de-cades by the unwillingness of the United States to act, let alone to lead. These wisdoms also demonstrate that many of the tools and techniques of the economic reform agenda can play a vital role within environmen-tal law and policy, but only if they are carefully restricted to their legiti-mate sphere of competency. Absent the moderating force of a competing language for identifying and evaluating our environmental obligations, the language of economic reform has begun to appear inescapable, even infallible. Yet this overextension, this implicit claim by economics to re-solve questions for which it was never devised, now threatens not only the fullness of environmental law but also the environmental economic discipline's own coherence. Rather than dismantle or displace the script, therefore, this book aims to restore balance to it by highlighting, expli-cating, and—hopefully in the process—safekeeping the endangered wis-doms of our environmental laws.

*

The story of economic reform typically begins with the advent of mod-ern U.S. environmental law, a period of unprecedented legislative activ-ism in the environmental, health, and safety arenas, which began when Congress passed the National Environmental Policy Act (NEPA) in the waning days of 1969. A variety of salient, culture-altering events, such as the first appearance of an image of the Earth from space, the success of Rachel Carson's _Silent Spring,_ the controversial attempt to dam the Grand Canyon, the pollution-induced burning of the Cuyahoga River, and Sen-ator Gaylord Nelson's organization of the first Earth Day, had fueled American environmentalism to the point that no politician could ignore the force of the movement. Thus, with extraordinary bipartisan zeal,

Congress passed, and the president signed into law, not only the NEPA but also the Clean Air Act (CAA); the Occupational Safety and Health Act; the Federal Insecticide, Fungicide and Rodenticide Act; the Marine Mammal Protection Act (MMPA); the Clean Water Act (CWA); the Endangered Species Act (ESA); the Toxic Substances Control Act; the Resource Conservation and Recovery Act; and the Comprehensive Environmental Response, Compensation, and Liability Act.[2]

Many of these early federal statutes mandate ambitious environmental, health, and safety goals with deliberate insensitivity to the costs of achieving them. The CAA, for instance, requires the Environmental Protection Agency (EPA) to set health-based national ambient air quality standards for certain criteria air pollutants without regard to cost. The Delaney Clause of the Food, Drug, and Cosmetic Act (FDCA) similarly bars the Food and Drug Administration (FDA) from approving any food or color additive for human food products that has been "found to induce cancer in man or animal," with no room for comparing the additive's risks against its benefits. Nor are humans the only species to receive the benefit of such rigid standards of protection. As interpreted by the Supreme Court in the famed case of *Tennessee Valley Authority v. Hill*, the ESA reflects Congress's belief that the value of endangered species is literally "incalculable" and that federal agencies must avoid taking any action that would impair the survival of such species irrespective of the benefits given up in order to do so. The CWA likewise imposes an anti-degradation requirement on "high quality" waters, meaning that states must preserve the quality of those waters regardless of how severely the duty requires them to limit effluent discharges.

In other areas, cost is included as a factor in setting environmental, health, and safety goals, but only as a way of establishing an outer limit for what level of protection the government can demand of regulated actors.[3] For instance, although the preamble to the CWA announces a goal of eliminating all pollutant discharges into navigable waters by 1985, many of the statute's operative provisions moderate this aspiration by requiring facilities to adopt only the best "available," "practicable," or "economically achievable" pollution control technology. Similarly, the Occupational Safety and Health Administration (OSHA) is empowered by its organic statute to establish regulations that will "assure so far as possible every working man and woman in the Nation safe and healthful

Examples of no-cost statutory regs.

working conditions." The actual regulatory standards promulgated by OSHA, however, take due account of the "feasibility" of achieving them, a term that permits the agency to consider how financially onerous compliance with its proposed regulations will be for industry and other regulated actors.[4]

These traditional regulatory approaches—standard based, technology based, and feasibility based—have proven successful in a variety of contexts, even according to the narrow criteria of success offered by the economic framework. The CAA, for instance, is consistently shown by retrospective cost-benefit analyses to have generated health benefits that vastly exceed the amounts expended by public and private actors to attain them.[5] Equally heralded in hindsight is the decision by the United States to phase out the use of certain ozone-depleting chemicals well in advance of robust empirical demonstration of their dangerous properties and to spearhead an international effort to require other nations to do the same. Notwithstanding such successes, a groundswell of opposition to the traditional approaches of environmental law arose around the time that Ronald Reagan took office as president in 1981 and has continued unabated to this day. Then and now, would-be reformers decry the absolutism and seeming naïveté of the early statutes and urge wholesale alteration of the nation's approach to environmental, health, and safety regulation. To be sure, these critics acknowledge, conventional regulatory approaches have been successful at plucking some low-hanging fruit. But, they argue, as the nation continues to seek ever greater levels of environmental quality and human safety, its focus must shift to a more balanced, economically oriented approach, lest it risk perverse, ineffectual, and potentially impoverishing consequences.[6]

what about criminal liability?

The case against conventional environmental, health, and safety regulation has proceeded on several fronts. With respect to specific design or technology requirements, reformers point out that mandating a single compliance technique might be easy to administer, but it might also preclude opportunities for companies to achieve the same level of protection through other, less costly means. Put more harshly, such "command-and-control" regulation repeats the great mistake of command economies: It fails to appreciate the inherent diversity, dynamism, and innovative capacity of competitive markets, believing instead that faithful bureaucrats can light upon just the right pollution control technique from the perch of their Washington offices.[7] Worse still, the approach might encourage a culture of complacency among regulated firms: Once firms have installed acceptable designs or technologies under a command-and-control regime, they seem to have no incentive to reduce pollution emissions further. An economic fine or tax on emissions, on the other hand, allows firms to experiment with finding the cheapest and most effective technology for reducing emissions *and* provides them with continuing incentives to eliminate pollutants right up to the vanishing point. Armed with such theoretical advantages, reformers argue that environmentalists—who still can be heard to criticize taxes and other economic regulatory tools for allowing companies to purchase a "right to pollute"—seem to value abstract principles more than actual improvements to air and water quality.

criticisms of tax-based policies ~ only value abstract principles? ✓

Uniform environmental or health quality standards also have come under attack, as reformers note that the cost of achieving a given standard might vary dramatically in different locations or for different firms. A more efficient approach, sensitive to the overall social welfare effects of regulation, would allow for variation in standards based on factors such as compliance cost, population density, individual preference, and so on. One means of affording such flexibility that has attracted significant attention among reformers is the tradable emissions permit, a policy instrument that also has come to dominate worldwide policy discussions regarding climate change. On this approach, rather than require each firm in a given industry to reduce pollution emissions to a uniform level, the regulating agency instead issues permits that equal in total amount the level of pollution desired after reductions. Firms able to reduce emissions at low cost then will have valuable, unused permits to trade to high-cost firms, ultimately leading to the same net environmental outcome at a much lower cost than uniform mandated reductions. Reformers also question the assumption that government agencies should seek to provide the same level of environmental quality or health protection for all citizens, given that control costs may vary across individuals, industries, or regions of the country. Why, for instance, should a municipal water facility in a small town be required to meet the same drinking water standards as a facility located in a large urban area, when the per capita costs of doing so might be orders of magnitude higher for the smaller facility? Because of uniform standards, reformers argue, the residents of the small town are forced to pay a higher per-unit price for drinking water than urban residents, whether or not they actually desire the higher level of water safety.

The critique of feasibility-based regulatory approaches extends this reasoning further. In practice, feasibility-based approaches tend to become a "knee of the cost curve" heuristic. That is, regulators tend to require greater and greater levels of environmental or human health and safety protection, up to the point that compliance costs begin to sharply rise. In the view of the economic reformers, this approach only gets things half right. It is simply not the case, the reformers argue, that more protection is always better. Nor is it the case that more protection is always better so long as it does not require great costs to achieve. Sometimes great costs are justified by even greater benefits. Conversely, sometimes small costs are unjustified because they generate even smaller benefits. The answer to the question "How safe is safe enough?" depends in every instance on a careful assessment of both the costs *and* the benefits of investing in safety. So too do investments in environmental protection. In both cases, the relevant task for the regulating agency is to assess the costs and benefits of various standards and to select the standard that, according to the agency's calculation, maximizes overall social welfare.

Of course, if costs and benefits are to be compared in this manner, the economic approach requires regulators to place monetary values on human life, endangered species, old-growth forests, and other protected subjects of environmental, health, and safety law. Many defenders of the traditional approach argue that this effort to commensurate life and money is simply indefensible, much like the effort—in their eyes illegitimate—to confer a "right to pollute." As the reformers point out, however, such commensuration occurs at least implicitly any time we draw a regulatory line at something less than absolute protection (which is to say, in practice, all the time). Moreover, in many instances, the regulator finds herself explicitly in an apples-to-apples situation. Influential studies published during the 1970s and 1980s appeared to demonstrate that safety regulations themselves can have unintended, risk-enhancing consequences. Laws that mandate seat-belt usage may encourage people to drive faster due to a sense of greater security, at least partially offsetting the safety benefits of wearing seat belts;

childproof-container requirements may cause parents to take less care in hiding medicine bottles from children because, again, the parents are lulled by the protection seemingly offered by the safety device.[8] In other instances, the product or activity under regulatory scrutiny may provide potential environmental or safety benefits, in addition to its suspected risks. Hence, genetically modified crop seeds may cause unpredictable ecological harms, but they may also lower pesticide use or reduce pressure to convert rainforest to agricultural land due to higher productivity levels. Similarly, strict scientific scrutiny of pharmaceutical drugs may help to prevent harmful products from reaching the market, but this scrutiny may also prevent ailing individuals in the interim from accessing treatments that turn out to be beneficial.

If the logic of cost-benefit balancing seems inescapable in such "risk-risk" situations—because the familiar distinction between life and money (or, relatedly, between the government's "doing" and "allowing" harm) has been deconstructed—then it becomes even more so as the risk-risk literature is supplemented by a "health-health" one.[9] Using widely observed correlations between higher levels of income and improved health and longevity, some commentators contend that the mere act of requiring expenditures on regulatory compliance creates adverse health consequences by displacing private income. These commentators claim to identify, at least in theory, an amount of regulatory cost that *necessarily* implies the loss of a statistical human life, in addition to whatever health benefits the regulation aims to achieve. Through such a radically reductionist lens, all of public regulation becomes life-versus-life, rather than life-versus-money or precaution-versus-profit, as environmentalists would have it. And although advocates of this commensurated worldview have yet to extend their approach to environmental harms, a facially plausible basis exists for doing so. Specifically, in light of empirical studies that purport to find a causal relationship between growth in per capita national gross domestic product (GDP) and environmental quality,[10] analysts might claim, again at least in theory, that regulatory expenditures reducing or delaying economic growth per capita necessarily imply a cost in terms of environmental quality. From this perspective, regulation not only entails an inherent health-health trade-off, but an environment-environment one as well.

Again, the logic of instrumentalist balancing seems inescapable in the face of such claims. Thus, there is little wonder that reformers depict their approach as the only "rational," "sensible," or "realistic" one for grappling with the challenges of risk regulation. They buttress their arguments further by pointing out that the cost-benefit-analysis approach relies on rigorous empirical assessment of environmental, health, and safety conditions, while the conventional U.S. environmental law approach seems too often to use vague or theoretical bases for concern as a moment to impose onerous regulations. The conventional approach, they argue, partakes of what is known more familiarly within international environmental law as the "precautionary principle."[11] Proponents of the precautionary approach to environmental decision making emphasize the limits of human knowledge and the frequency of unpleasant surprises from technology and industrial development; thus, they advocate an ex-ante governmental stance of precaution whenever a proposed activity meets some threshold possibility of causing severe harm to human health or the environment. One oft-cited articulation of the precautionary principle, for instance, seeks to trigger a process of public scrutiny and regulation through the admonition, "When an activity raises threats of harm to human health or the environment, precautionary measures should be taken even if some cause-and-effect relationships are not fully established scientifically."[12] Under this approach, regulators are to adopt measures that are proportionate to the threat perceived and that are open to revision as knowledge develops, but they are not to be hampered by a default assumption against government regulation in advance of complete scientific demonstration of harm.[13]

Although rarely invoked by name, the precautionary principle does find expression in U.S. environmental laws, agency interpretations, and judicial decisions adopted during the 1970s. As Judge Skelly Wright observed in an important early CAA case, "Where a statute is precautionary in nature, the evidence difficult to come by, uncertain, or conflicting because it is on the frontiers of scientific knowledge, the regulations designed to protect the public health, and the decision that of an expert administrator, we will not demand rigorous step-by-step proof of cause and effect. Such proof may be impossible to obtain if the precautionary purpose of the statute is to be served."[14] In the view of economic reformers, this precautionary orientation leads to a variety of ills. Too often, it seems, courts and regulators are willing to prohibit activities, award compensation, or require technological changes based only on speculative evidence that a threat of harm exists. Importantly, the reformers argue, such conservatism does not merely risk unnecessary compliance costs or deprivation of economic opportunities; it also brings into play all of the risk-risk scenarios described above, in which life and the environment are unintentionally sacrificed in the name of saving life and the environment. In that respect, the precautionary principle seems to offer an asymmetric concern for the consequences of government action; it focuses on the harms to be prevented by regulation without considering the harms caused by regulation, including especially regulation's forgone benefits. According to the precautionary principle's detractors, therefore, the principle either must be expanded to include an obligation to consider the environmental, health, and safety costs of regulatory activity—in which case it would become woefully indeterminate and need to be replaced by something more comprehensive, such as cost-benefit analysis—or the principle must be rejected as a one-sided tool that is likely to prolong a range of harms that would be alleviated in its absence.

In light of arguments such as these, numerous commentators within U.S. academic and policy-making circles have concluded that the normative case in favor of the economic approach to regulation is simply overwhelming and that competing approaches such as those associated with the precautionary principle are either incoherent or inappropriate as frameworks for risk regulation. The conclusion of these thinkers is particularly notable in light of their acknowledgment that economic tools such as cost-benefit analysis suffer from a number of conceptual and practical limitations of their own. Nevertheless, proponents of the economic approach increasingly believe that the "first-generation" debate about the approach's normative desirability is over and that today the important questions concern "second-generation" issues regarding how best to implement cost-benefit analysis in the environmental, health, and safety regulation context.[15] The basic superiority of the economic approach to environmental, health, and safety regulation, in other words, is no longer seriously doubted. And not only, they argue, must this realization be consolidated domestically, so that U.S. environmental

[margin handwritten note: Precautionary principle must include costs of regulation?]

law is reformed to shed its precautionary origins, but it also must be spread internationally, so that other nations begin to rest their environmental policies on the twin foundations of [economic analysis] and [sound science.]

*

This is only a breezy and incomplete version of the economic reform story. Many of the arguments just summarized are revisited in greater detail and with more care in later chapters. The point in this introduction is simply to convey a sense of the script's tenor and to argue that it has become sufficiently pervasive to crowd out long-familiar alternative languages for conceiving and confronting environmental problems. In early 2007, the White House Office of Management and Budget (OMB) provided an even breezier version of the script when discussing what it calls "social regulation," such as environmental, health, and safety rules. OMB's rendition purports to provide an airtight depiction of how regulation *in fact* should be conducted: "Regulation that utilizes performance standards rather than design standards or uses market-oriented approaches rather than direct controls is often more cost-effective because it enlists competitive pressures for social purposes. . . . Regulation that is based on solid economic analysis and sound science is also more likely to provide greater benefits to society at less cost than regulation that is not. Thus a smarter or better regulation program relies on sound analysis and utilizes competition to improve economic growth and individual well-being."[16] The aim of this book, again, is to show the incompleteness of this line of reasoning, to highlight certain ways in which the very ubiquity of the script has masked its gaps and inconsistencies. Beneath omnipresent and seemingly innocuous invocations of "economic analysis and sound science" lie enormously challenging moral and political quandaries—quandaries that not only are inadequately resolved by the economic paradigm but also are beclouded by the paradigm's very presence.

The task of exposing these limitations—and of remembering environmental law—requires addressing up front what seems to be the most indefensible aspect of the traditional, precautionary approach to environmental law and policy: the apparent inattention demonstrated by the precautionary principle to the costs of environmental, health, and safety regulation, including costs that *themselves* take the form of environmental,

OMB "script"

health, and safety harms. Despite the seemingly unimpeachable logic of the economic critique of the precautionary principle, the role of actions and opportunities, of "doing" and "allowing" harm, actually turns out to be much more complicated and interesting than the reformers' account reveals. Undoubtedly, reformers are correct to note that no society should flatly ignore forgone benefits and other negative effects of precautionary regulation. But this is a trivial observation, for no serious proponent of the precautionary principle disagrees with it. Despite frequent caricaturing of the precautionary principle as a crudely absolutist doctrine, its proponents actually regard the principle as merely one aspect of a much more elaborate regulatory process in which the precautionary principle is applied with a view toward proportionality of response and adaptability over time.[17]

Thus, just as no physician would unthinkingly and universally follow the precautionary mandate of the Hippocratic adage—putting down injection needles and refusing to reset broken bones out of literalist adherence to the "First, do no harm" injunction—no regulator would adhere to the precautionary principle without paying attention to forgone benefits, unintended consequences, new information, and changed circumstances. As this book argues, what appears in both cases to be an emphasis on the dangers of acting (or doing) to the disregard of "opportunity costs" actually represents a much more subtle acknowledgment of the irreducibly distinctive nature of human agency, of the unique relationship of authorship that the doer bears to the done. At bottom, the precautionary principle does not command political communities to do this or that thing on behalf of the environment, any more than the time-honored words of Hippocrates issue specific behavioral directives to physicians on behalf of patients. Rather, the precautionary principle asks political communities to simply remain mindful of the special position from which they choose and act. It asks them to remember that they stand in a relationship of responsibility not only to their own citizens but also to other states, other generations, and other forms of life, all of which—like the physician's patient—reside in a condition of at least partial dependency on the decisions and actions of the regulating body. The precautionary principle asks the political community to maintain such ethical self-awareness through the principle's subtle reminder that doing harm, even in service of a greater benefit, is something that the community must always *own* as part of its identity.

As it turns out, there is a deep metaethical proposition at stake in this staged contest between the traditional precautionary approach to regulation and the economic optimization approach. To expose this proposition requires at the outset a seeming detour into the realm of moral philosophy on the individual level. Thus, this book begins by showing that the normative ethical theory of utilitarianism—which commands individuals to choose actions that result in the greatest net-positive consequential impact, much as cost-benefit analysis asks policy makers to maximize the overall welfare impacts of regulation—suffers from a fundamental conceptual difficulty, in that its command of impartial causal optimization serves to erode the very basis on which individuals have come to think of themselves as agents whose choices and actions somehow matter. By attaching no weight or significance to the relation of ownership that individuals hold with respect to their choices and their effects, the utilitarian rubric forces individuals to treat their actions as mundane, as essentially indistinguishable from all other elements of the complex causal order, including the indefinitely many omissions (or opportunity costs) that are created at every temporal instant of their lives' narratives. As critics have pointed out, it is difficult, within such a homogenized conception of the causal order, to imagine how over time individuals would maintain a strong sense that their actions merit the disciplinary attention of ethical norms of any sort. Instead, the utilitarian normative standard seems to transform the individual into simply a part of the furniture of consequential optimization. Such a standard may be well and good for actual pieces of furniture, but for conscious, reasoning human beings it is difficult to perceive how, in the long run, these subjects would abide such a denial of their own distinctiveness; accordingly, it is difficult to see why they would continue to want to *be moral,* however such a notion is defined. And once the foundational desire to be moral is lost, so too must be the goal of optimization, along with all other normative ethical standards of behavior.

Thus, rather than sublimate human agency to a standard of causal optimality, individuals instead must continue to navigate the world in pursuit of their own visions of human flourishing, attempting to do

[margin annotations:]
Society must own the harms they create...

critique of Utilitarianism ✓

individuals want to continue to be moral...

good, while plagued at every instant by limited information, limited control, and limited assurance of success for their chosen projects. They must recognize that no standard of normative ethics, however persuasively theorized, can eliminate the fundamental obligation to *independently* evaluate the variety of reasons that exist for choosing and acting at any given moment and, indeed, to be ready at any moment to generate reasons for choosing and acting that are uniquely their own. No decision-making rubric can permit the individual to ventriloquize morality.

Whether this conception of agency at the level of individual ethics has bearing on the conception that we must or should adopt at the level of collective human behavior is a complicated question. Obviously, there are important differences between individual human subjects and collective human institutions, such as regulatory agencies or nation-states. This book nevertheless argues that the analysis just rehearsed can be translated into the collective domain and that such translation serves to give meaning and coherence to the traditional precautionary orientation of environmental law. Like utilitarianism for individuals, the welfare-maximization approach to regulation tends to suggest that government policies are "hostage to what the facts turn out to show in particular domains,"[18] so that no distinctive notion of collective discretion and responsibility is deemed necessary or appropriate for the fashioning of public policy. Thus conceived, however, the approach is unable to account for the normativity of what the facts tell us—that is, for the assumption that some agent somewhere should act in accordance with the facts so discovered. Put differently, an applied welfare-maximization approach such as cost-benefit analysis is unable to account for its own prescriptivity.[19] Where, after all, do we locate the agent that has independently reasoned to the point of accepting welfare maximization as a standard of choice, as opposed to simply stumbling upon it, with the label "maximal" somehow already affixed? We cannot say that welfare maximization is desirable simply because it maximizes overall welfare; that derivation assumes the regulating community already has accepted maximization as its fundamental desideratum. Where then do we locate the *sense* that maximization is compelled?

This sense must reside with some collectivity—a political community, for instance—that has established norms and procedures for self-governance and that has selected welfare maximization as a goal using those mechanisms. Perhaps this is where we find ourselves now, reaching as we are toward a "cost-benefit state,"[20] in which welfare maximization implemented through regulatory cost-benefit analysis provides our *übernorm* for public policy making. The rub, however, is that this calculus of choice is premised on the notion that public policy should impartially and objectively reflect the determinants of individual well-being, paying no heed whatsoever to goals or interests that are articulated at the collective level. The approach seeks precisely to eliminate collective discretion and judgment by formalizing and determining—empirically— the content of public policy according to *individual* welfare consequences. Thus, to be complete, our embrace of welfare maximization must also be fatal to the very notion of *our* embrace. Perhaps for this reason, welfare economists periodically slide into nonindividualistic argumentation. As will be seen in later chapters, they often report that society "consents" to particular levels of risk based on valuation evidence inferred from blue-collar-labor markets or that future generations will be "better off" for having some of their members lives sacrificed in furtherance of welfare maximization. Such rhetorical slides into collective or group-based thinking reflect the fact that much of the subject matter of environmental, health, and safety regulation simply resists figuration within the thinned-out lens of methodological individualism.

[margin note: No attention to goals at a collective level]

[margin note: rhetorical slides into group-based thinking]

In the long run, this book argues, the welfare-maximization approach not only proves disruptive to the project of reasoning through certain daunting collective issues, such as international and intergenerational environmental responsibility, but also undermines even its own attractiveness as a standard of social choice. As with impartial optimization and its agency-eroding effect on the individual level, the mandate of welfare maximization is curiously self-undermining. Proponents of the economic reform project argue that a rational political community should recognize the various pathologies that infect the exercise of its collective agency—bureaucratic excesses, interest-group machinations, moments of populist hysteria—and conclude that its collective self-identity is best expressed in the form of self-negation through formalized welfare analysis. Despite the surface appeal of such a cost-benefit state, the case for surrendering collective responsibility to a program of optimal decision making lacks one essential element, namely, an account of how a political community can maintain conviction in a standard of social choice that implicitly denigrates the community's judgment and, indeed, denies the community's very existence. Before entering finally into this cost-benefit state, we should demand from the economic reformers a persuasive account of how our community will continue to cohere, once we have given up the language within which we recognize and reshape ourselves.

language with which we reshape ourselves

*

With more detail than the preceding discussion, part 1 demonstrates that a political community must always, in a nontrivial sense, stand outside of its tools of policy assessment, maintaining a degree of self-awareness and self-criticality regarding the manner in which its agency is exercised. The precautionary principle encourages such conscientiousness by reminding the political community, poised on the verge of a policy choice with potentially serious or irreversible environmental consequences, that its actions matter, that they belong uniquely to the community and will form a part of its narrative history and identity, helping to underwrite its standing in the community of communities, which includes other states, other generations, and other forms of life. Like the Hippocratic adage for physicians, the precautionary principle reminds the cautioned agent that life is precious, that actions are irreversible, and that responsibility is unavoidable. Such considerations, in contrast, hold no clear or secure place within the logic of welfare maximization, tending, as it does, to deny the political community a view from within itself and to ask the community, in essence, to regulate from nowhere.[21]

Precautionary principle "reminds"

Part 2 begins to outline the ethical and political questions that cannot be asked, let alone answered, within the language of the economic paradigm and that therefore underscore the necessity of maintaining collective engagement with environmental law's normative dimensions. It examines specifically the perils of prediction, showing that the tools of risk assessment and cost-benefit analysis—foundational elements of the economic approach to environmental law and policy—inevitably leave promissory notes in the form of ineliminable empirical uncertainties, irresolvable valuation questions, and other nontechnocratic issues that implicate collective responsibility. A case study of the hurricane-prevention planning process supervised by the United States Army Corps of Engineers for the city of New Orleans following Hurricane Betsy in 1965 is used to demonstrate the normatively inflected nature of modeling assumptions and other discretionary judgments that lie at the foundation of risk assessment. It is shown that, although Army Corps representatives repeatedly depicted the hurricane-levee planning process as leading to an "optimal" level of storm-surge protection for New Orleans, their claim was ill supported even before the tragedy of Hurricane Katrina in 2005.

Part 2 also interrogates various aspects of the economic approach to predicting welfare effects of policy choices, including how it constructs the dichotomy between efficiency and equity, how it assigns monetary values to human life and the environment, and how it grapples with the emergence of values and meanings within the operation of policy mechanisms themselves, including those mechanisms that purport merely to tabulate individually inscribed interests. It will be argued that practitioners of cost-benefit analysis often adopt quite contestable positions from among the range that one could take on these foundational modeling decisions. Worse, the standard practices of the profession then tend to crowd out other ways of conceptualizing well-being and promoting its attainment; significant moral and political questions become effectively embargoed, as the optimizing logic of cost-benefit analysis seems to attach an efficiency "price" to any deviation from *its* norms of evaluation. Rather than debating what versions of welfarism could best be deployed in furtherance of public policy, citizens are invited only to inspect whether government agencies are maximizing the use of their tax dollars according to unexamined rules and techniques of valuation.

Upon close inspection, then, both risk assessment and cost-benefit analysis—two essential methodologies offered in support of the claim that public policy making can be reduced to empirical technique—reveal a need for the political community to maintain a skeptical distance from its tools of understanding and evaluation.[22] In the absence of this distance, questions concerning how the community should behave in the face of uncertain but potentially irreversible or catastrophic threats become obscured by technical assumptions and assessments that do not merit the degree of deference afforded to them. The political community likewise becomes tempted by the notion that its regulations and policy pronouncements can simply trace individual preferences, without also endogenously affecting them. The downside of such a cabined vision becomes evident in the context of powerful new technologies, such as genetic engineering or nanomedicine, and dramatic shared experiences, such as Hurricane Katrina or the emerging effects of climate change. In such potentially culture-altering cases it is inadequate to hinge policy making exclusively on our existing preferences, given that those preferences are either ill formed or nonexistent and, in any event, will be altered significantly by the very outcome of our policy making.

Rational
choice?

Cocaine/ta
example?

Part 3 turns to the vexing question of how a political community should regard non-nationals, future generations, and other interest holders who are not granted full membership in the cost-benefit community. This set of questions requires more than simply enhancement of our empirical understanding of natural and economic systems; it requires mechanisms for fostering improved democratic dialogue and developing new collective norms of responsibility. How, for instance, should we evaluate the interests of human individuals who live in different political communities, yet whose well-being is deeply affected by our collective choices? Should they be counted on an even footing with "our" citizens for purposes of welfare maximization? Should the statistical value of their lives be reduced to reflect lower per capita income, as is often the case with economic analyses of climate change? Should their welfare be counted at all? We cannot answer these questions through the use of cost-benefit analysis, because the procedure requires some analytically prior determination to be made regarding who is entitled to membership in the cost-benefit policy-making community and on what basis their interests are to be counted. In theory, of course, utilitarian and welfarist philosophies have a cosmopolitan bent, in that the reasons behind their commitment to equal regard for individuals are difficult to confine to those individuals who happen to be members of a given political community. Yet, when translated into applied methodologies such as cost-benefit analysis, these philosophies must be deployed by policy makers who serve as agents for a particular, confined community. Thus, whether and how to evaluate the extraterritorial impacts of policies re-

Extra-territorial
impacts...

main open questions that cannot be resolved from within the cost-benefit methodology itself.

Similar difficulties plague the evaluation of effects on future human generations and other forms of life. The economic approach to regulation typically values such interest holders only insofar as they are held as objects of value by the presently living. That approach, however, is shown in part 3 to be inadequate to guarantee a sustainable environmental context capable of supporting life over time. In order to truly respect the interests of future generations, we must undertake an engaged effort to anticipate and consider the details of their plight and to provide the specific institutions and resources they will need in order to endure it. This is one of our greatest political challenges: Because the needs and interests of future generations are not self-presented in the manner typically demanded of liberal subjects, we must develop an ethic of care for an unknowable other. This normative challenge, already humbling, will become even more fraught when we consider other forms of life—for which we have even less innate affinity to guide us—and potential forms of life, such as genetically engineered persons, over which our powers of design and control will stress our ethical resources to a degree rarely seen in human history. Obviously, simple environmental maxims such as the precautionary principle do not provide sufficient guidance concerning how best to address such issues; they do, however, underscore the need for the development of new norms of international, intergenerational, and interspecies responsibility. The economic approach, in contrast, offers the implicit and misleading message that our needs consist only of better data and more-rigorous techniques of valuation.[23]

Finally, part 4 reevaluates contemporary debates over how best to pursue environmental law going forward. Through a detailed examination of recent efforts by the EPA to implement portions of the CWA using both cost-benefit analysis and more-traditional regulatory approaches, part 4 shows that the latter display numerous virtues that have become underappreciated. Most notably, precautionary approaches can be defended as being particularly well suited to safeguarding life and the environment under conditions of uncertainty and ignorance, as opposed to the conditions of probabilistic sophistication that are presupposed by proponents of the economic approach. Contrary to prominent critiques, the precautionary approach to environmental regulation does not require agencies "to be universally precautionary"[24]; accordingly, the approach does not actually entertain the problems of inconsistency and paradox that critics ascribe to it. Instead, the precautionary approach focuses on particular categories of harm and separates them out for special treatment within the larger "hydraulics" of biophysical and sociolegal systems. Viewed sympathetically, this asymmetry of concern represents

a procedurally rational attempt to catalyze empirical investigation, redress political imbalance, and respond with prudence to threats of a
potentially catastrophic or irreversible nature. In many real-world contexts, heuristic decision making of this nature embodies what leading
cognitive psychologists have called "ecological rationality"[25]—that is,
pragmatic decision making that is both well tailored to the informational
and cognitive constraints of actual choice environments and capable of
evolving and adapting over time.

*ecological
rationality (?)*

The precautionary approach to environmental law also embodies
an underappreciated *expressive* wisdom. Even if mechanical devices of
the kind sought by economic proponents could be identified to grapple
with uncertainty, to resolve analytically prior ethical questions such as
wealth and resource equity, and to address challenges posed by future
generations and other unknowable others—in short, even if economic
proponents could fully formalize environmental lawmaking—the approach still would fail to register the sense of regret that accompanies
risk regulation's tragic choices. Risk-risk, health-health, and environment-
environment tradeoffs may be in some sense inevitable, as the economist reminds us, but they are *regrettably* so. Thus, although the aspirational
standards of the precautionary approach do entail a shortfall between
statutory command and societal achievement, such shortfalls serve to
remind the political community that it must remain constantly engaged
in searching inspection of how to create a society in which tragic
choices are not so starkly and pervasively posed. Lives lost under the "Do
the best we can" approach of traditional environmental laws are not
viewed as efficient tradeoffs, blithely accepted in exchange for whatever
welfare has been gained. Instead, they are viewed as tragic, lamentable
consequences of human fallibility and finitude—a moral remainder that
provides enduring motivation for surviving members of society to seek
ways of doing better in the future. In contrast, because the economic approach aims to express a formal, comprehensive rationality, it must invariably round this moral remainder to zero.

Part 4 concludes by bringing together the book's various critiques
and themes within a framework of environmental constitutionalism. In
earlier chapters, features of the dominant analytical approach to environmental law are shown to have a similar, problematic form: They
treat that which should be outcome determining as, instead, outcome

determined. The likelihood that uncertain but potentially catastrophic events will occur is made to depend on the rough subjective estimates of technocrats, when instead the likelihood should hinge on whether the responsible political community is willing to countenance disaster among its legacy of achievements. The monetary worth of human life is made to reflect workers' apparent willingness to pay to reduce their occupational risk, when instead the value of those workers' lives should be enhanced through democratically conferred legal protections. The environmental resources of future generations are made to depend on the present market rate of interest, when instead that market rate of interest should arise from a legal backdrop that places ecologically determined sustainability constraints on resource use. As part 4 reveals, an environmental constitutionalism, in which certain needs and interests of present and future generations, the global community, and other forms of life are given foundational legal importance, would help restore conceptual coherence and priority to the subjects of environmental law. Again, the wisdom of earlier environmental approaches becomes evident: Although not formally enacted as constitutional amendments, many landmark environmental statutes can be seen as efforts to exert a foundationalist impact. Indeed, they can be seen as visionary attempts— perhaps not yet failed—to attain the kind of constitutional prominence that Bruce Ackerman has identified within federal civil rights statutes, as part of his probing analysis of how America's constitutional culture has continued to thrive during the twentieth century *beyond* the much-studied arenas of formal amendment and judicial decision making.[26] Accordingly, to help reawaken the constitutional vision archived in our landmark environmental laws, this book closes with a model statute, the Environmental Possibilities Act, which would restructure agency decision making to comport better with the environmental wisdoms we once knew.

*

In strong terms, critics have attempted to dismiss the precautionary principle as "puzzling," "incoherent," "indeterminate," "paralyzing," "worse than unhelpful," and "literally senseless."[27] These critiques, however, have overlooked the most desirable feature of the precautionary principle, which is not necessarily the level of environmental, health, and safety harm that it promises to avoid, but rather the more subtle manner in which the principle reflects and reinforces a notion of the political community as a distinct entity with special responsibility to evaluate its decisions and actions in the context of other states, other human generations, and other forms of life. Despite its seemingly unequivocal command, the Hippocratic adage "First, do no harm" is not only or even primarily a behavioral prescription. It is instead a subtle but steadfast reminder to the professional so cautioned that his or her actions carry distinctive moral weight and responsibility. It is a reminder most fundamentally to *be moral*. Similarly, the precautionary principle's requirement that we pause to consider the environmental consequences of our actions is at bottom a reminder that social choices express a collective moral identity —*our* identity, an identity that cannot be located within the freestanding optimization logic of economics, although we need to consider its content more now than perhaps ever before.

Motivational — express, evolve morality

Reminding — constantly look at ethical notions

Endogeneity — preferences evolve from regulatory choices

International/"other" — what to include?
    how to include them?

PART ONE REGULATING FROM NOWHERE

CHAPTER ONE

Agency and Optimality

AMONG THE MANY MORBID THOUGHT EXPERIMENTS to appear within philosophical writing over the past century, perhaps none is more engrossing and perplexing than Judith Jarvis Thomson's exploration of the Trolley Problem.[1] Its setup is deceptively simple: Imagine yourself the engineer of a train barreling steadily down a set of tracks toward a group of five unlucky souls who will, with certainty, be killed unless you decide to divert the train to a side track immediately. The only catch: stuck on the side track is a single person who will, again with certainty, be killed if you decide to spare the five by switching tracks. Neither the five nor the one have become stuck on the train tracks through any fault of their own, and none of them have otherwise engaged in any wrong-doing that would merit punishment. All have lives that they personally regard as worthy of continuation. Your first question, then, is whether it is morally permissible to switch tracks in this situation.

Now imagine yourself a surgeon confronted by a patient who will, with certainty, die unless you perform an operation on him before the end of the day. The only catch: you have five other patients in your waiting room who will, again with certainty, expire before the end of the day unless you perform organ transplant operations to save their lives. All of your patients are morally innocent and desire to continue living. Because this is a purely hypothetical experiment, we can enrich the situation by making you the world's most skilled transplant surgeon and by making your patients lie in need of lung, kidney, and heart transplants. Thus, if you would only harvest these organs from your first patient, you could with certainty save the lives of the five patients in your waiting room. Hence, your second question: Is it morally permissible to sacrifice the one to save the five in this situation?

With remarkable consistency, individuals tend to believe that it is permissible to sacrifice the one to save the five in the first scenario but not in the second. The obvious and confounding issue then becomes why? If a kind of consequentialist-utilitarian reasoning—which counsels choosing the course of conduct that will lead to the best state of affairs according to overall utility or well-being—holds sway in the first scenario, why not in the second? When are we permitted to take life in order to save life, and when are we not? A great deal of ink has been spilled on these and related questions, with no hint of consensus yet emerging. The most we seem to be able to say is that the scenarios trigger a cluster of inarticulate notions that individuals hold regarding the concept of agency, notions that are not easily accommodated within utilitarianism but that are so basic and omnipresent that at least one scholar has suggested they reflect a "universal moral grammar."[2]

Most obviously, the scenarios tap into a long-controversial but stubbornly enduring conceptual distinction between acting and omitting (or between doing and allowing, killing and letting die).[3] Something about the former class of behaviors—acting, doing, killing—seems much more centrally associated with the exercise of one's agency, and therefore more morally culpable, than the latter—omitting, allowing, letting die. And something about the organ transplant scenario seems to require a much greater degree of affirmative behavior on the part of the surgeon than is required of the engineer in the train scenario. By switching tracks, the engineer is thought to merely permit the one to die, while the physician, by harvesting the organs, quite actively causes the death of the one. This act-omission distinction often is said to underlie the precautionary principle, the Hippocratic adage, and other such seemingly asymmetric moral imperatives (to wit: "First, *do* no harm"). It also finds expression in the law, for absent prior conduct, a special relationship, or some other exceptional circumstance, individuals in the Anglo-American legal tradition generally are under no duty to aid others, even if they

Surgeon vs. train example [handwritten margin note]

could save the life of an imperiled individual at little or no cost to themselves.[4]

Although it clearly violates the utilitarian injunction to maximize the overall impact of one's causal capacity, this act-omission distinction appears with surprising consistency in the reactions of individual subjects to psychological experiments designed to test their moral-reasoning processes. Researchers have shown, for instance, that parents seem willing to accept a greater risk of harm to their child if it occurs passively in the form of a failure to vaccinate against a risk of disease than if it occurs actively in the form of a side effect from a vaccination that the parents ordered.[5] Some psychologists have taken to referring to such behavior as reflective of an "act-omission bias,"[6] given the manner in which the behavior seems clearly misguided from the perspective of optimizing health outcomes alone. Some have even proposed educational interventions during early childhood to encourage more comprehensive and calculative thinking by lay individuals, an approach that the psychologists believe ultimately will lead to an improved democratic citizenry.[7]

Because utilitarians hinge the normativity of behavior on its expected consequences for overall utility or well-being, they join these psychologists in criticizing the act-omission distinction. From the utilitarian perspective, the act-omission distinction seems to sanction indifference in the face of widespread suffering simply because the suffering does not register as having been caused by an individual's affirmative action. For instance, both Peter Singer, in his influential essay "Famine, Affluence, and Morality," and Peter Unger, in his evocatively titled book *Living High and Letting Die*, derive tremendous mileage from the simple empirical fact that most individuals who live in wealthy nations could, at little sacrifice to themselves, save dozens of lives each year through monetary donations to effective international aid organizations such as UNICEF or Oxfam. In their view, widespread embrace of utilitarian ethical theory would have the immediate salutary effect of requiring substantial transfers of wealth to serve the pressing needs of the global poor. Without the illusory comfort of the act-omission distinction, they argue, our choices would become radically more other-regarding.

Critics also point to failings in the concepts of causation and action that seem to be driving the act-omission distinction. A greater degree of ethical imagination, they contend, would lead individuals to see that occasions of suffering often are not outside the individuals' sphere of influence, but rather are well bound up with the exercise of their agency through intricately interrelated causal webs. Like complex, adaptive biophysical systems, sociolegal systems such as markets exhibit characteristics of interconnection, feedback, nonlinearity, and emergence. Disaggregating causal influence within such a "complicated tissue of events"[8] is no small task, as the myriad components comprising the system are both necessary and by themselves insufficient to account for the system's condition at any given moment. From this perspective, it is a cognitive failing—and not a morally relevant distinction—that we do not perceive the variety of ways in which our economic and political arrangements are inextricably intertwined with conditions of suffering throughout the world, even if we typically cannot point to a discrete, affirmative act of harm on our part.[9]

Nevertheless, something about our views on the normativity of behavior remains resistant to a simple reduction to "the numbers," to an equation of rectitude in behavior with exactitude in calculation. In the train-switching scenario, the numbers do seem to matter, at least to the extent that the engineer can with ethical permission divert the train from the path of the five to the one. In the transplant scenario, on the other hand, the fact that five lives could be saved by sacrificing the one seems utterly beside the point. One prominent alternative explanation for this distinction focuses on the degree of "necessariness" that appears to be involved in the sacrifice of the one within the two scenarios. We can imagine, after all, an alternative world in which the side train track does not place anyone at risk, because no one would happen to be stuck on the track at the pertinent moment. In the transplant scenario, on the other hand, the death of the one seems to be a necessary element of the plan. That is, we can scarcely imagine an alternative unfolding of events that saves the five but that does not require the purely instrumental sacrifice of another life-in-being. In that sense, the organ transplant scenario requires of the actor an apparently greater degree of purposefulness in the taking of life than does the train-switching scenario. In the latter context, the taking of life seems only to be an unintended, albeit predictable, secondary effect of saving the five.

Thomson offers a variant of the hypotheticals designed to deal with this double-effect explanation. If we imagine that the train track actually

[margin handwritten note: Lack of an alternative to [illegible] which no harm abscribes?]

forms a loop at the point of decision, rather than splitting into two separate tracks that continue on indefinitely, then we can further imagine that the train would double-back on whichever side of the track is not chosen by the engineer, in the absence of some obstruction to slow its movement. Thus, we now face the complication that hitting the one or the five *is* instrumentally necessary to slow the train down enough to stop it from doubling-back and hitting the individual(s) on the track not chosen. Even in this variant, Thomson argues, most people continue to believe that it is morally permissible to choose the death of the one rather than to let the train follow its present course into the five. Thus, instrumental use of the one seems not to inhibit maximizing the number of lives saved. Compare that intuition, however, with still another variant in which saving the five requires pushing a large individual, who happens to be standing over the tracks on a footbridge, down onto the tracks in order to slow the train adequately to avert collision with the five. In this scenario, individuals tend to react as they do in the transplant scenario: The action, they believe, is *not* morally permissible. But why?

[The loop] and [the footbridge] variants also cast doubt on another explanation sometimes offered for our divergent reactions, this one focusing on the special professional role of the physician. Both as an ethical and as a legal matter, we imbue certain professions with elevated duties of care and fidelity. The surgeon accordingly owes a duty to the patient that would be grossly violated by the nonconsensual harvesting of organs for transplant into other individuals. The action would be, in a real sense, the ultimate breach of trust that a physician can perpetrate against a patient. The train engineer, although a professional with public-safety obligations in some important respects, is not similarly positioned in a relationship of direct responsibility toward the individuals on the train tracks. Accordingly, in the initial train hypothetical, switching tracks does not seem to involve a breach of trust owed by the engineer specifically to the individual who happens to be stuck on the side track, at least not akin to the duty of fidelity owed by the physician to her patient. Such a notion of professional fidelity plays an important role in the story being developed in this book, in the sense that much of the ethics of professionalism is designed to promote self-consciousness regarding the distinctive role played by physicians, attorneys, and other agents who are

[handwritten margin note: No breach of trust]

entrusted with matters of fundamental importance to their clients. Nevertheless, as the loop and the footbridge variants of Thomson's hypotheticals demonstrate, professionalism does not fully account for the divergent reactions that individuals have to the hypothetical situations.

Still another explanation seeks to interpret individuals' reactions as being consistent with utilitarianism after all. On this account, individuals are thought to be projecting a nonarticulated understanding that, although certain courses of harm-causing conduct seem to be compelled in particular circumstances by utilitarianism, the actions would have longer-term or systemic adverse consequences that outweigh the situation-specific gains to be had. For instance, although harvesting the organs of the one to save the five in the transplant scenario may initially seem utility-maximizing, the broader consequences of such an action may include an erosion of public trust in the medical profession and an overall decline in social well-being. Similarly, a practice of pushing strangers off of footbridges may have antisocial effects in a way that a train engineer's momentary decision to switch tracks would not. With reasoning along these lines, many theorists have argued that the relevant empirical question to govern conduct under utilitarianism is not whether a particular act in any given circumstance would promote overall utility, but whether a general rule of behaving in certain ways across circumstances would do so.[10]

Expanded
utilitarian
arguments?

The problem with these expanded utilitarian arguments is that they have a stylized, almost ad hoc quality about them. Commentators seem to be shaping their empirical understanding of the consequences in order to fit what they believe to be the right thing to do—precisely the opposite of the analytical approach that utilitarianism claims as its defining feature. Moreover, the commentators still have not accounted for why certain seemingly utility-maximizing actions (for example, pushing strangers off of footbridges) would have antisocial effects, while others (for example, switching trains to a track where one individual is imperiled rather than five) would not. It seems that their story about the long-term social consequences of behavior depends on individuals in the world continuing to accept a distinction between kinds of behavior that the utilitarian framework rejects. In this respect, there is a danger, often cited in theory and sometimes witnessed in practice, that utilitarianism would be positioned as a framework for evaluating behavior that only the elite can handle, while the masses utilize more crude "moral heuristics" that are tolerated by utilitarians only because they approximate utility-maximization in their long-term effects.[11] Accordingly, when the heuristics of the masses fail significantly to achieve utility-maximization, they are to be treated as cognitive "mistakes" by the utilitarian elite, subject to "correction" through educational or regulatory interventions. Such a re-description of the felt moral experience of large portions of humanity may give utilitarian theorists comfort in the face of widespread apparent rejection of their approach, but it fails to pay adequate respect to the *reasons* that individuals have for rejecting the utilitarian approach.

Debiasing?

*

How, then, do we pay respect to the reasons individuals have for rejecting pure utilitarian thinking? How do we render sensible the felt moral experience that what we do is somehow more intimately bound up with our identity and our narrative history than what we fail to do? A number of important arguments have been offered, most of them proceeding from the standpoint of assuming, hypothetically, the utilitarian's duty of impartial causal optimization and pointing out undesirable or untenable aspects of such an approach. For instance, as Bernard Williams famously argued, attaching some special significance to the affirmative expressions of an individual's agency enables the person to be "identified with his actions as flowing from projects and attitudes which in some cases he takes seriously at the deepest level, as what his life is about."[12] In contrast, by fixing the normativity of behavior on a wholly impartial assessment of causal opportunities, utilitarianism renders the individual's projects and goals perpetually contingent, perpetually at risk of being trumped by alternative courses of action that happen to generate more expected utility. Most disruptively, these alternative courses of action may become obligatory, based on empirical features of the world that the individual has done nothing to choose or promote. Thus, even granting the claim that we are far more deeply imbricated in webs of causation and suffering than we typically acknowledge, the extreme position of utilitarianism—which seems to abandon entirely the notion of the individual's "separateness" for purposes of prescribing conduct—is both psychologically untenable and philosophically problematic. As Williams argued, the extreme position constitutes to the individual, "in the most literal sense, an attack on his integrity,"[13] a profound impairment of the ability to act and choose in a manner that reflects and reinforces the individual's own deeply held values and life pursuits.

Other critics have observed that the duty of impartiality imposed by utilitarianism tends to render infeasible the types of human connections (for example, bonds of kinship, love, and friendship) that we regard as significantly constitutive of our lives. Those relationships often depend, at bottom, on some degree of unabashed partisanship, yet the duty of impartial causal optimality does not permit the individual to express such partisanship, to allow, in other words, *some* to have a greater claim on the individual's time and effort than *all*.[14] Again, the utilitarian approach seems to render vulnerable some of the most sacred aspects of the individual's life. Versions of utilitarianism that would allow the individual to weigh more heavily the needs and interests of loved ones do not eliminate this concern, for the individual's ability to promise fidelity to another is still disrupted. The promise remains forever contingent on whether the empirical facts happen to permit the promise to be fulfilled; hence, it becomes not a promise at all. Little wonder, then, that our familiar relations of intimacy and allegiance seem difficult to square with utilitarianism: Although we undoubtedly agree with Singer and Unger that our conduct should become more cosmopolitan and other-regarding, the framework of utilitarianism seems to demand something far more extreme, an impartiality that would, in effect, homogenize our social world.

Still others have asserted the basic impossibility of obtaining the impartial, objective empirical assessment that utilitarianism seems to demand of the individual. Part of being ethical, no doubt, does involve an effort to identify what would produce the best consequences, all things considered, on an impartial and objective basis. Yet we also must acknowledge that the effort to occupy this "view from nowhere"[15] never will be fully successful. Our subjectivity always reasserts itself; we are always, at a minimum, conscious of the fact that whatever objective view is held represents *our* particular effort to achieve an objective view. Nor would we necessarily want the objective view to become our exclusive, seamless vantage point, given the degree to which our subjective perspective, our own unique point of view, forms an essential part of who we think we are. Much of the explanation for the attractiveness of agent-relative theories of individual ethics, which permit the normativity of behavior to hinge at least in part on considerations that arise from the individual agent's own particularity, lies in the ability of those theories to prevent our lives and our identities from cognitively receding into the broader causal systems within which we are situated. Again, ecology and other emerging scientific understandings of complex, adaptive systems underscore the importance of this function, given that they posit an expansive yet simultaneously imperfect causal potential on the part of any single component within such systems. The challenge for the individual of crafting an ethical identity under these conditions is especially profound: Not only does the individual face innumerable opportunities to act, but he also experiences his actions as deeply embedded within a causal order that belies the classical belief in predictable dyadic causal relations.

Although seemingly ignored by utilitarians, it is precisely this tragic condition that lends a melancholy cast to the category of opportunity costs: We must act, and by acting we must at every moment create forgone options that later might haunt us. Under a duty of impartial optimization, this "saddest" category—what "might have been"[16]—becomes much more than a reflective indication of the challenge of crafting an identity in a world of indefinitely many imperfect causal opportunities. It becomes a reflexive imperative to act and to choose in a manner that draws no boundaries between the human actor and the complex causal system within which she is situated.[17] Again, it is difficult to see how individuals would abide such a scheme: The approach of impartial utilitarianism seems to unduly flatten the world, forcing the individual to allow her projects, her relationships, and her very life to be rendered indistinct within a commensurated and ultimately barren landscape.

Or, at least, so say a great many moral philosophers. A growing band of neuroscientists, psychologists, and legal commentators, however, contend that the moral philosophers' rejection of utilitarianism is simply an elaborate analytical effort to craft an "ought" from an "is," that is, an uncritical desire to lend philosophical pedigree to the common-sense intuitions of individuals, to offer, in essence, an "exercise in moral *rationalization*" rather than in actual "moral *reasoning*."[18] To be sure, individuals do appear to employ an ethical system that includes deontological constraints and other nonconsequentialist features, as psychological experiments demonstrate. But, according to the skeptical observers, this evidence may have very little to do with the work of moral philosophy; instead, it may reflect biologically determined cognitive processes, which moral philosophy simply seeks to redeem through ex-post explanation. As psychologist Joshua Greene states, "According to this view, the moral philosophies of Kant, Mill, and others are just the explicit tips of large, mostly implicit, psychological icebergs. If that is correct, then philosophers may not really know what they're dealing with when they trade in consequentialist and deontological moral theories, and we may have to do some science to find out."[19] Some of this requisite science is being conducted using neuroimaging techniques, which map the brain activity of individuals as they deliberate over the Trolley Problem and other philosophical thought experiments. From these studies, a picture is emerging of "higher" and "lower" order brain functioning, in which those individuals whose choices reflect utilitarian reasoning tend to exhibit greater levels of activity in parts of the brain associated with planning, analysis, calculation, and other complex cognitive functions. In turn, individuals whose choices reflect deontological constraints, such as the "First, do no harm" injunction, tend to exhibit greater levels of activity in parts of the brain associated with emotion, intuition, affect, and other comparatively simple cognitive functions. On this account, divergent reactions to the Trolley Problem and other hypotheticals are seen as biologically determined: Emotionally charged hypotheticals, such as the footbridge and the organ harvesting scenarios, place us in the grasp of our lower-order, intuitive brain system, while more abstract hypotheticals, such as the original train-switching scenario, permit our higher-order, calculative system to hold sway. The upshot for deontologists is that their philosophical project has been reduced to an apologia for evolutionarily determined beliefs and behaviors. To Greene, moral philosophy is merely epiphenomenonal of biology: "[W]hat we find when we explore the psychological causes of characteristically deontological judgments might suggest that what deontological moral philosophy really is, what it is *essentially,* is an attempt to produce rational justifications for emotionally driven moral judgments, and not an attempt to reach moral conclusions on the basis of moral reasoning."[20]

A recent brain-imaging study lends further support to this view by helping to demonstrate that activity in certain regions of the brain actually appears to cause moral judgment, as opposed to resulting from such judgment.[21] Researchers studied patients who had suffered damage to the ventromedial prefrontal cortex, a region of the brain associated with the generation of emotions. Although both the intellect and the baseline mood of these individuals were unaffected by their brain injuries, the individuals did display diminished responses to emotionally charged pictures and generally lower levels of "social emotions," such as embarrassment, guilt, and empathy. By comparing these individuals' reactions to moral philosophical hypotheticals with those of nonimpaired individuals and individuals who suffered other kinds of brain injuries, researchers were able to demonstrate significant behavioral differences among the individuals with diminished social emotional capacity. Specifically, these individuals tended to behave similarly to the other groups when they were confronted with abstract or impersonal moral dilemmas, such as the original train-switching hypothetical. However, when they were confronted with more personal, emotionally freighted dilemmas, such as the organ harvesting hypothetical or the footbridge variant of the train-switching hypothetical, these individuals continued to focus on "the numbers," displaying a much greater willingness than other subjects to sacrifice the one in order to save the five. In the researchers' terms, the brain impairment of the studied individuals appeared in these contexts to "produce an abnormally 'utilitarian' pattern of judgements."[22]

To be clear, when these scientists describe "an abnormally 'utilitarian' pattern of judgements," they are not intending to weigh in on the merits of competing ethical theories. They are simply identifying a departure from the "norm" of nonutilitarian decision making. But many commentators *do* regard these studies as carrying an implicit evaluative message. After all, utilitarianism has been associated with the higher-order cognitive functions of planning and analysis and with regions of the brain that are "among those most dramatically expanded in humans compared with other primates."[23] Deontological judgments, in turn, have been linked with the lower-order functions of affect and emotion and with regions of the brain that are thought to be evolutionarily older and more primitive. Thus, while commentators acknowledge that emotionally driven judgments are often useful—because "emotions are very reliable, quick, and efficient responses to recurring situations, whereas reasoning is unreliable, slow, and inefficient in such contexts"[24]—they contend nonetheless that in many contexts our emotional responses

Type I
vs.
Type II ?

lead us astray. We should recognize, they argue, that the deontological tradition arises not from a full-fledged philosophical framework, but from the use of "moral heuristics . . . or rules of thumb, that work well most of the time, but that also systematically misfire."[25] Indeed, they contend that the deontological prohibition on knowingly taking innocent human life may misfire in just this fashion: Although our moral squeamishness over sacrificing innocent life has a deep-rooted evolutionary explanation and may well serve valuable purposes in some contexts, in other contexts—such as when "the deaths are relatively few and an unintended by-product of generally desirable activity"[26]—it may be an obstacle to clear thinking.

To sum up, then, a growing number of neuroscientists, psychologists, and legal scholars have adopted the view that utilitarianism, deontology, and other normative ethical systems are best understood not as philosophical theories to guide behavior, but as ex-post accounts that seek to depict biologically determined cognitive processes as having flowed from autonomously selected philosophical theories. Prescriptively, then, we should seek to better understand the emotional bases of our moral judgments and to guard against them in contexts, such as environmental, health, and safety decision making, where the emotional resonance of the task is likely to impede our thinking. To be clear, these moral skeptics are not saying that we should suppress our capacity for social emotion altogether. That capacity, after all, serves a wealth of purposes that complement, rather than are substituted by, the higher-order brain functions of planning and calculation. But they do express mounting concern that our hard-wired capacity for compassion and empathy— our ability to treat every life-in-being as sacred—is actually ill suited to grapple with the demands of our contemporary situation. In some situations, such as those involving the inevitable tradeoffs of life-and-death decision making, and for some individuals, such as those who are government regulators vested with responsibility over tragic tradeoff situations, the emotional system associated with deontological judgment should be seen as something *to be overcome* rather than as something to be accounted for, much less relied upon.

Emotional system as something to be overcome(?)

*

Like the economic-reform narrative, this narrative of overcoming moral philosophy should be approached cautiously. After all, there is a certain irony in the fact that moral philosophy's own thought experiments are being used to generate empirical evidence of the field's supposed obsolescence. A further irony, barely worth pointing out, lies in the fact that only those individuals suffering diminished capacity for generating social emotions—that is, only those suffering a form of brain damage—appear capable of displaying the elevated adherence to utilitarianism that commentators seek from public policy makers. A final irony, this one well worth pointing out, emerges from the commentators' simultaneous embrace and denial of biological determinism. On the one hand, human judgment is depicted as flowing more or less automatically from perception and cognition. We are said to reject saving the five in the organ transplant and the footbridge hypotheticals not because those situations are philosophically distinguishable from the train-switching hypothetical, but because they simply "code" to us as more emotionally immediate and personal, thereby activating a different region of brain function. On the other hand, the commentators' prescriptive advice seems to be driven by a sense that we can, at least over time, learn to "recode" certain situations, so that they get channeled to regions of the brain that are associated with higher-order cognitive function. One wonders, though, why such reasoned efforts to guide our thinking only become operative, as opposed to epiphenomenal, when the scientist himself moves from the domain of observation to prescription.

This is not the place to debate determinism, but even within the commentators' own advice, there seems to be a recognition that philosophy and reason have some role to play beyond mere ex-post rationalization of our immutable patterns of cognition. Why, after all, do the train-switching and footbridge hypotheticals come to us "tagged" with different affects, so that they get channeled to different centers of brain functioning? Why is one scenario usually considered abstract and impersonal, while the other tends to be felt as immediate and emotionally resonant? For skeptics of philosophy, the explanation for this difference will take the form of an evolutionary story regarding the traditional closeness of contact within human groups and the ensuing need for bright-line, hard-wired injunctions against doing harm. But the story also will include talk of cohesion, cooperation, and culture.[27] Because philosophy is an integral part of culture, it seems quite premature to conclude that the discipline has nothing to offer in the understanding and guidance of our conduct, that its work is hermetically sealed off from human evolution and the functioning of the brain. Conversely, the field of neuroscience seems too young, the brain too complex, and the relationship between culture and cognition too interwoven to believe that we can reduce ethics to a scientific enterprise, to believe, for instance, that we can evaluate ethical reasons according to their evolutionary status within a purported hierarchy of brain functioning or that we can neatly extricate those aspects of "social emotion" that have outlived their usefulness from those that have not.[28] We simply do not know what all might be lost from such an equation or from such an extrication.

As evidence for these propositions, we might consider one of the subjects in the brain-imaging study described above, an individual who, after finding himself consistently willing to sacrifice lives in order to promote overall well-being in the hypotheticals, stated to a researcher with apparent surprise, "Jeez, I've become a killer."[29] We might also consider—against the advice of the biological determinists—one final thought experiment from moral philosophy, this one concerning the incompatibility between utilitarianism and the basic architecture of a conception of ethical agency. This thought experiment will not only help us make sense of the subject's surprise at having assumed the view from nowhere and "become a killer," it also will point the way toward a deeper understanding of the value of the precautionary principle within environmental, health, and safety law.

Imagine, for the moment, that utilitarians are correct that our behavior should be directed toward the optimal use of our causal influence, treating all agents impartially and attaching no special significance to our affirmative actions, to those choices and behaviors that we might otherwise be inclined to shroud with narrative continuity and attach to something we call our life. Imagine that we abide this notion and that we no longer distinguish our actions from the indefinitely many omissions that are inevitably created at every temporal moment of our existence. Imagine, in short, that we sublimate ourselves to the *übernorm* of impartial causal optimality. Where, then, do we locate the normativity of optimality? That is, where do we locate the sense that causal optimality is ethically compelled?

We may join the skeptics of philosophy and simply respond that this sense is located in the dorsolateral prefrontal cortex or the inferior parietal lobe. But we ought to be able to say more than this about the grounds of our ethical standards. The skeptics assume rather uncritically that the normative superiority of different cognitive and behavioral patterns can be fixed by reference to an evolutionary concept of sophistication. But it will always remain open to us to interrogate that which is said to reflect our "higher" order capacities. These interrogations themselves may be biologically determined, as the skeptics would suggest, but they also may be part of a more complicated and reason-affirming process of co-evolution between culture and cognition.[30] The choice of conception is largely ours and, in truth, no one seems to be wholly or consistently determinist. As noted above, something enduring about the notion of ethical agency and the possibility of human reason seems to be revealed by the very fact that skeptics of philosophy urge the adoption of policies and institutions to promote greater use of the capacity for utilitarian judgment.[31]

A fundamental metaethical proposition can be glimpsed here. As Samuel Scheffler has explained, a normative standard of instrumental optimality for individuals—which attaches no special significance to the viewpoint, actions, or other features of the agent herself—must derive its attraction, paradoxically, from "considerations other than instrumental optimality."[32] The contrary notion that instrumental optimality has primary normative significance faces insuperable complications. For instance, as argued later in this book, there may be circumstances in which the deliberate attempt to optimize may not be the most sure route to optimization. Knowing when and how to depart from the optimization calculus in favor of more pragmatically sensible approaches therefore implies the existence of some independent agent responding to at least some additional normative criteria. Philosophers who favor a utilitarianism of rules, rather than of acts, face a similar challenge in that they must offer an account of how one derives the appropriate degree of generality or abstraction to apply in the selection of rules. That is, lest we invite an infinite regress, some agent somewhere must reach outside of a purely utilitarian calculus in order to establish the pitch of her rule-utilitarianism.

Even more fundamentally, as Scheffler argues, the desirability of holding oneself to a norm of instrumental optimality cannot be premised on a judgment that it is the instrumentally optimal thing to do, because "if it did, one would need already to have accepted the norm in order to see oneself as having reason to accept it, which means that the proposed derivation is circular."[33] To avoid such circularity, one first must posit an independent human subject who views herself as peculiarly responsible for the affirmative expressions of her ethical agency, in the sense that those expressions are guided by reasons that she herself has considered and chosen. That very brand of discretion, however, seems to be what utilitarianism, through its systematized decision-making procedure, eliminates from our moral reasoning.

Scheffler's argument bears some relationship to legal philosopher Daniel Markovits's effort to construct an account of "the necessary architecture of the first person," in which Markovits argues that any meaningful conception of personal ethical agency must include a recognition of oneself not only as an agent responsive to reasons for acting, but also as a generator of reasons, including reasons that are intimately and uniquely one's own.[34] While Scheffler argues that the mere fact that individuals view themselves as subject to ethical norms of any sort implies that they must accept a distinction between their agency and the larger causal order, Markovits argues that a minimal logical requirement of individuals being able to coherently view themselves as ethical agents is an ability to supply reasons for acting that are not solely dictated by an external normative theory, such as the optimization rubric underlying utilitarianism. From either perspective, the rub is that by urging a standard of pure impartiality and programmatic decision making, causal optimizers also implicitly ask individuals to deny the belief that their judgments and actions are ethically distinctive—the very belief that seems to be a minimally necessary precondition for having reason to accept any theory of normative ethics.

Defenders of optimization might respond that the conceptual separateness of ethical agents is maintained in their framework by the fact that the utilitarian calculus takes account of an individual's own particular causal position and information set when assessing optimal courses of action. In that sense, the individual's particularity enters in as the baseline from which the costs and benefits of conduct are estimated. Optimizers, however, have no way of cabining their logic, for presumably individuals also should choose to position themselves within causal settings and to invest in obtaining information in a manner that is calculated to achieve optimal outcomes. Soon enough, the duty of causal optimality becomes infinitely self-referential, and the individual becomes lost within a framework that achieves its goal of consistent ethical treatment only by denying the basis on which individuals have come to think of themselves as distinctive ethical agents.

Indeed, the optimization framework of utilitarianism is not only unhinged—in the sense that it exogenizes the process by which its intended audience develops and maintains a sense of personal urgency concerning the framework's subject matter—but it also is expressed in a formal language that implicitly condemns the discretion and judgment of its subjects. Like any theory of normative ethics, utilitarianism seemingly must depend for its relevance and coherence on the existence of agents who are empowered to respond to reasons for acting, including at least some reasons for acting that are entirely independent of an externally imposed normative framework. Unlike most other theories of normative ethics, however, the formal language of optimization disparages any such independent reasons for acting. That is, the formalized ethical world of optimization offers the individual a series of stark choices: partiality versus optimality, subjectivity versus rationality, and, as we shall see in later chapters, equity versus efficiency, and precaution versus maximization. Thus, although the optimization framework depends for its persuasiveness on the continued self-awareness and cognitive independence of the agents it seeks to persuade, the framework's axiomatic structure simultaneously and unavoidably condemns those agents' independent judgments as leading to suboptimal outcomes. Under such a conception, it is hard to imagine why individuals would maintain sufficient investment in the exercise of their agency to continue the quest for ethical behavior at all. Indeed, after reflexively following the mandate of optimality long enough, they might simply wake one day to report to themselves, "Jeez, I've become a killer"—and not have an account of why.

condemns independent judgments as suboptimal

*

One can crudely schematize much of the preceding discussion along the lines of figure 1, which offers a spectrum of normative ethical theories arranged by the level of "thickness" or "thinness" accorded to the individual agent within the theory. "Thickness" or "thinness" in this context means the degree to which the particularity of the individual agent—her beliefs, her discretionary judgments, her chosen projects, her personal

thickness + thinness

affiliations, and so on—are permitted to enter into the determination of behavior's normativity. To be clear, this is *not* intended to describe a spectrum of self-interested (or egoistic) and selfless (or altruistic) norms of behavior, although many commentators quite quickly run these categories together. Rather, the spectrum captures something more in the nature of degrees of subjectivity or objectivity: To what extent, in other words, do the individual's choices and actions under a given ethical theory emerge from reasons that she has selected or fashioned, as opposed to being fixed according to externally given criteria? Thomas Nagel uses the terminology of "agent-neutral" and "agent-relative" reasons for acting to capture a similar notion. As he describes, "If a reason can be given a general form which does not include an essential reference to the person who has it, it is an agent-neutral reason. . . . If on the other hand, the general form of a reason does include an essential reference to the person who has it then it is an agent-relative reason."[35] Accordingly, a range of positions might be identified between the extremes of pure agent-relativity, which would permit the normativity of behavior to rest at all times on reasons that emerge solely from the individual agent, and pure agent-neutrality, which would require the individual at all times to conform her decision making to externally given criteria of rightness or goodness—again, bearing in mind that agent-relativity in this sense need not strictly correlate with self-interested, as opposed to self-determined, behavior.

Proceeding from the extreme of pure agent-neutrality, the individual in some versions of utilitarianism is given a slighter "thicker" presence, in that the individual is allowed to have interests or preferences that are subjectively determined, rather than externally given. Indeed, as the next chapter explains, welfare economics tends to insist that the grounds of value be located exclusively in the expressed or revealed preferences of individuals. Still, the utilitarian approach remains largely agent-neutral, in that the individual is not permitted to weigh her interests in a different manner than the interests of others or, more broadly, to reason independently to the normativity of behavior. Subjective interests or preferences enter into ethical decision making only as empirical inputs, and decision making itself remains formalized and impartial. Forms of moderate utilitarianism tend to relax this calculative mandate in ways that afford the individual agent a larger scope for self-determination. For

Pure agent-neutrality	Utilitarianism	Moderate utilitarianism	Moderate deontology	Deontology	Pure agent-relativity

FIGURE 1. A Spectrum of Agent-Relativity in Normative Ethics

instance, [rule-utilitarianism] can be thought of as slightly more agent-relative than act-utilitarianism, in that the theory implicitly acknowledges the cognitive and informational demands placed upon the individual by act-utilitarianism, substituting in its stead a less taxing regime of general behavioral axioms. Thus, rather than devote her mental energy to sifting through the indefinitely many alternative courses of conduct that she could engage in at any moment (many of which no doubt would promote overall well-being to a greater extent than her currently planned course), the individual instead may use the mental space freed up by rule-utilitarianism to pursue projects and to follow reasons that are her own.

Proceeding from the other extreme of pure agent-relativity, one encounters first the vague but essential category of [deontology] which often is associated with the act-omission distinction, given that deontological constraints characteristically focus on·what the individual does rather than on what she fails to do. In its strongest form, deontology prohibits certain actions as immoral—killing, lying, stealing, and so on—irrespective of whether they may lead to desirable consequences and, indeed, irrespective of whether they may prevent a greater amount of the very type of harm that is prohibited by the deontological decree. Such constraints are relative to the particular ethical agent: What matters is not whether she minimizes killing, but whether *she* kills. Forms of moderate deontology relax this constraint in various ways. For instance, when faced with an opportunity to eliminate killing above a certain threshold—say, five lives saved for one death caused—the individual may be given permission to engage in the otherwise prohibited action. This approach remains distinct from the utilitarian approach of causal optimality due to the critical role played by the [agent's discretion.] As

[margin annotations:] deontology, strict proscription of certain acts

relaxed through an agent's discretion

Scheffler has written, "[t]he *permission* not to produce the best states of affairs suffices to free individual agents from the demands of impersonal optimality, and thus to prevent them from becoming slaves of the impersonal standpoint."[36]

To some readers, deontology might seem to be just the opposite of agent-relativity, in that it often is associated with abstract reasoning and rules that are largely insensitive to context. Although such deontological approaches do impose obvious and important constraints on behavior, they still afford wide latitude to the individual to pursue projects of her own choosing and to resolve conflicts between duties according to reasons that she determines persuasive. The deontological principal of beneficence, for instance, obligates the individual to improve the life conditions of others, but not in the reflexive sense of utilitarianism's duty of impartial causal optimality. The agent remains a reflective and reasoning force, her discretionary judgment operative in a nontrivial sense with respect to the effectuation of the duty. Indeed, in the classic Kantian formulation, even the strictest prohibition against killing is agent-relative in the sense that the prohibition exists as a result of the individual agent having reasoned herself to the point of accepting it as a categorical imperative. Moreover, the individual at all times not only regards herself as a "thick" reasoning agent in this manner, but she regards all other human individuals as three-dimensional agents as well. Rather than being required simply to identify and maximize the interests of others, the deontological individual is required to recognize and respect the irreducibly distinctive agency of others—greeting them not as objects of empirical inspection but as subjects of reasoned discourse. As chapter 5 argues, this difference of perspective takes on profound importance when translated into the realm of global environmental politics, in which states must engage their sovereign counterparts in continuing efforts to reason toward shared goals rather than simply measure and accept the behavior of other states as empirical givens when fashioning environmental law and policy.

The central question raised by this discussion is not one concerning where, on the spectrum of agent-relativity, the most attractive normative ethical theory lies. Rather, it concerns how we can continue even debating where that theory lies, once we have embraced the utilitarian denial of the distinctiveness of ethical agency. That is, unless one subscribes to the biological determinist viewpoint, in which the impartial optimality of utilitarianism *just will* become the highest expression of our cognitive capacity, some degree of agent-relativity seems unavoidable, for without it we no longer can locate the individual who regards her choices as sufficiently distinctive to merit ethical inspection of any sort. To be sure, as we move along the spectrum away from pure agent-relativity, we may feel something like a sense of moral progress, as the latter-day critics of deontology claim. For instance, we initially seem to have been well served by "primitive" deontological constraints, in that they rendered certain violent or antisocial behaviors absolutely forbidden in an easy-to-follow, taboolike fashion. But, the story continues, as we thought more deeply about the needs and interests of others, we came to recognize that our causal capacity could be even more other-directed. That is, our negative categorical duties to avoid harm could be supplemented by positive duties to relieve suffering. Eventually, overcoming our moral squeamishness, we came to recognize that we might even *cause* certain harms with justification, if the actions are necessary to avoid the imposition of much greater harms.

At this point, we stumbled across an enticing notion: Perhaps ethical behavior could be perfectly described in an optimific calculus, perhaps the impartial maximization of interests could be fully realized in the world, if only we would submit our decision making to utilitarianism's comprehensive rationality. Here, though, we encounter a difficulty: Once fully inside the "view from nowhere" that utilitarianism demands of us, we no longer remember why it was that we felt obliged to behave *that way*. Our steadfast effort to strip normative ethics of its agent-relativity leads us to a dramatic, possibly irreversible, confusion: We no longer remember why our behavior matters, because we no longer have a mechanism for ascribing that behavior peculiarly to us. The unique relation of composition that an individual holds to the use of her causal capacity fades from view, along with all other such "primitive" ethical notions. Moving steadily across the spectrum of agent-relativity toward more formalized, "objective" approaches, we reach our impartialist ideal, only to find the ground given way beneath us, our ideal immediately lost along with it.

[margin handwritten note: Cannot abscribe behavior to one particular individual, so no longer know why it matters?]

CHAPTER TWO

Prescription and Precaution

EARLY DOMESTIC AND INTERNATIONAL EFFORTS to eliminate [ozone-depleting substances] were based largely on theoretical arguments as to their potential for harm, a classic example of precautionary regulation in the face of incomplete information regarding potentially disastrous environmental harms. Years later, empirical investigations have confirmed the grounds of the scientific community's concerns, and cost-benefit analyses now are capable of "verifying" the wisdom of the early proactive stance on ozone. Significantly, though, precautionary wisdom emerged at the time of that action from a political body that saw itself as standing outside of, and being critically disposed toward, its tools of risk assessment and welfare maximization. Indeed, at the time that the United States led the global effort to reduce the use of ozone-depleting substances, computer programs were rejecting satellite data on the extent of loss in the ozone layer as being too far from the range of expected results to be valid.[1] Nevertheless, given the monumental potential harms at stake, the United States accepted present economic costs in order to eliminate the possibility of an ozone catastrophe and eventually convinced the rest of the industrialized world to follow its precautionary lead.[2]

Today, the precautionary approach is derided by U.S. policy elites as a "mythical concept . . . like a unicorn."[3] They note that the precautionary principle seems insensitive to the reality that every decision to act or refrain from acting implies a range of alternatives that, for better or worse, are not selected. For critics, it is these seemingly underappreciated consequences of behaving according to the precautionary principle's dictates that render the device objectionable as a basis for public policy making. Behaving with precaution entails its own costs, both direct (for example, compliance costs) and indirect (for example, opportunity costs, adverse side effects), which must be considered alongside whatever risk is being guarded against through precautionary action. For this reason, environmental law reformers argue in favor of a decision-making framework, such as cost-benefit analysis, that maintains fidelity to the assessment of consequences comprehensively, in pursuit of optimal outcomes. In turn, the reformers strongly criticize heuristics, such as the precautionary principle, which may turn out in individual cases to forgo substantial gains.

In raising these objections, critics of the precautionary principle seem to assume that the only meaningful way in which to think about potential hazards is to compare their likelihood and potential magnitude against the costs of averting them, in other words, to evaluate the consequences of government action or inaction. Hence, much of the debate concerning the precautionary principle tracks the fundamental divide in moral philosophy between consequentialist-utilitarian and deontological approaches to individual normative ethics. From this perspective, resorting to the precautionary approach does appear to be an abandonment of the quest to maximize an overall desideratum, such as utility or welfare, in light of the expected consequences of behavior. As the previous chapter argued, however, such optimific quests do not exhaust the universe of respectable and widely held theories of ethical behavior. Nor, at least with respect to individuals, are they necessarily the best-suited theories to regulate decision making when faced with conditions of radical uncertainty and limited powers of control. Under those daunting conditions, as critics of utilitarianism have argued, individuals suffer both a lack of sufficient raw material to satisfy the maximization mandate and a frustrated ability to imbue their lives with a sense of authorship and personal efficacy. Accordingly, an agent-relative approach to ethical decision making—that is, an approach that affords conceptual significance to the individual's particularity, in terms of her point of view, her judgment, her reasons—is desirable both in order to provide additional sources of traction within empirically ambiguous contexts and in order to ground a sense of abiding personal investment in the outcomes of ethical decision making.

If this metaethical analysis can be translated from the individual to the collective context, it will lend a great deal of concrete sense to the precautionary approach: Even granting the causal optimizer's claim that "risks are on all sides of social situations,"[4] that fact alone will not compel the adoption of an optimization standard in which risks imposed and opportunities forgone are treated as analytically indistinguishable. As shown in the previous chapter, such a homogenized conception of the causal order threatens to undermine the basis on which human agents have come to think of their actions as especially deserving of deliberation, choice, and responsibility. Put differently, no coherent conception of agency—perhaps even a collective one—can fully deny the distinctiveness of the agent's choices and actions in the manner compelled by impartial optimization. Instead, something like the admonitions of Hippocrates and the precautionary principle may be necessary simply for the implicit reminder contained within them, that the agent's actions and decisions carry normative weight that must be borne by the agent alone.

*

As with the degree of agent-relativity present in theories of individual normative ethics, one can construct a spectrum of normative political theories arrayed by the degree of "thickness" or "thinness" accorded within them to the state or other collective political entity (fig. 2). In this context, the notion of agent-relativity refers to the collective as an agent and captures the degree to which that entity, acting through its political institutions, is permitted and empowered to identify goals, pursue projects, and articulate reasons that emanate from the collective itself, as distinct from the atomistic individuals comprising the collective. To what extent, in other words, does the theory of political organization allow the normativity of government action to be defined with reference to the polity *as a polity*, separate and apart from the agglomerated interests of individuals that make up the polity?

The "thickest" approach to state power and discretion in this sense is totalitarianism, which describes a regime in which the state regulates

| Social choice theory | Libertarianism | Social liberalism | Communitarianism | Totalitarianism |

FIGURE 2. A Spectrum of Agent-Relativity in Political Theory

nearly every aspect of public and private behavior according to its own ideology. Whether of fascist or communist variety, the totalitarian state is hardly constrained in any way from engaging in actions or adopting policies on account of the status or interests of others, including the state's own individual members; instead, the limit of the state's authority emerges only from the possibility of revolution. In some respects, totalitarianism is the collective equivalent of pure agent-relativity for individuals, since in both cases the relevant agent pursues its own espoused life plan without constraint. And just as some philosophers have argued that pure agent-relativity effectively leads individuals to promote—and not necessarily to undermine—the well-being of others, some political actors have argued that an all-encompassing or total state leads governments to serve—and not necessarily to suppress—the well-being of society. The latter have just been far less persuasive than the former.

At the other extreme are various forms of social-choice theory, in which the state essentially disappears altogether and is replaced by a nonagentic apparatus for aggregating individual votes, preferences, or interests into a collective decision-making matrix. Although such procedures encounter well-known paradoxes,[5] their formal reduction of collective choice to individual interest aggregation represents perhaps the fullest expression of the liberal conviction that a just, efficient, and sustainable society can be attained while maintaining state neutrality among competing conceptions of the good. The state from this perspective becomes akin to the purely agent-neutral conception of normative ethics on the individual level, since the state's choices and actions are entirely fixed according to an externally given, formalized decision-making rubric. When juxtaposed against totalitarianism, the theoretical attraction of such an "anorexic" state is obvious: With perfectly described and transparently implemented rules for identifying, weighting, and aggregating preferences, the individual no longer needs to concern himself with the oppressive potential that seems to lurk within other, more self-consciously collective conceptions of political organization.[6] On the other hand, as various liberal theorists have pointed out, the individual may face a different, more subtle kind of oppression under optimific theories such as utilitarianism; namely, he may find himself conscripted into the project of maximizing overall well-being.[7] He may be told, for instance, that his talents and resources are more productively deployed in directions other than his chosen life course; that the utility generated by his group's religious practices does not outweigh the gains to be had from the majority group's promulgation of an official state religion; or, more generally, that his thick "separateness" as a person need not be respected within the cost-benefit state, since his thin "individualness" as an interest holder has already been acknowledged in the process of welfare assessment. Seen in this light, the attempt to perfect individualism for normative political theory ends up instead endorsing a form of totalitarianism.

The most important practical application of social-choice theory is welfare economics, which seeks to fix laws and policies according to a utilitarian calculus that remains characteristically focused on individually inscribed interests. Welfare economics underlies the environmental law reformer's effort to replace traditional policy approaches with those grounded in risk assessment and cost-benefit analysis. From this perspective, in order to answer the question "How safe is safe enough?" we need only tabulate the benefits to individuals from investments in safety and compare the result against estimated costs. Proponents recognize that this approach requires controversial techniques for valuing all potential sources of well-being according to a common metric and that it usually does not address the fairness of whatever distribution of well-being results from government policies. Still, due to a variety of pathologies that are believed to afflict the conventional approach to democratic lawmaking—such as the disproportionate influence exerted by special interests on the content of legislation, the capture of regulatory agencies by those same interests, and the failure of public understanding regarding complicated scientific issues—these reformers contend that society is best served by submitting its agency to a formalized assessment in which individual interests are aggregated into a collective welfare function.

Between the extremes of a totalitarian state and a state that is pro-grammed only to replicate the outcomes of an idealized, stateless order of individuals, one finds the related categories of libertarianism, liberal-ism, and communitarianism. They differ, again, in the degree to which they allow expression of the ideas, aims, or values of the political com-munity to occur along a nonindividualistic axis. For instance, the libertarianism, or "minarchy," of theorists such as Robert Nozick first identifies certain natural rights that are said to be held by citizens and then justifies the existence of the state as a separate entity only to the extent necessary to protect individuals' natural rights.[8] The liberties of individuals are negative, not positive; that is, the state is obliged to avoid interfering with individual autonomy (for instance, by taking private property or by impairing the obligations of contracts), but is not required to provide affirmative conditions that might be necessary for human flourishing (for instance, by guaranteeing health care or by redistribut-ing wealth). Here, then, one finds a version of the act-omission distinc-tion reified to the collective level. Under libertarianism, the state's "active" influence is described against an assumed and normatively unchal-lenged backdrop of private relations, a sort of prelegal or prepolitical so-cial order. The state "harms" on this account only by interfering with the private functioning of the market and other consensual social relations. If suffering, deprivation, and inequality do exist in society, such harms are not the fault or responsibility of the state, any more than such condi-tions are the fault or responsibility of individuals who have not "caused" the harms.

Just as utilitarians have critiqued the [act-omission distinction] for failing to acknowledge the inordinate ways in which individuals' omis-sions are not as clear-cut or benign as they imagine, critics have attacked libertarians for overlooking the flaws in their conception of a prelegal or prepolitical social order.[9] To these critics, the state is always actively en-gaged in the construction of the market and other sites of social order-ing. For instance, even the most minimal libertarian regime of property and contract law requires resolution of normative issues concerning the priority of conflicting property uses or the level of information and au-tonomy necessary for contract enforceability. Thus, in light of the un-avoidable degree of public influence that exists over even those spheres that are regarded as "private" in libertarian or classical liberal thought, adherents of social liberalism are much more willing to allow the state to regulate the use of property, to police the fairness of contracts, and to ensure the availability of health insurance, welfare, and other threads of the social safety net. As with liberalism generally, the focus of social liberalism remains very much fixed on the individual; thus, government authority remains limited by prohibitions on interfering with the freedom of expression, the pursuit of religious affiliation, and so on. Moreover, the ultimate purpose of government is not to structure the collective articulation and endorsement of particular values, goals, or ways of living, but rather to enable individuals to select among them, free from coercion or suppression and equipped with capacities for reason, self-determination, and tolerance that have been promoted through liberal education.[10] Nevertheless, because human autonomy and flourishing are seen to require that individuals hold certain positive liberties—such as the right to health care and economic opportunity—and because these positive liberties in turn are thought to create affirmative duties to act on the part of the government, the state in general appears "thicker" within social liberalism than classical liberalism.

Positive liberties...

Thicker still is the communitarian state, which follows from a critique of liberalism and its focus on the individual.[11] Communitarians regard the liberal conception of individual autonomy as somewhat contrived; in their view, liberalism often describes human nature and existence in an asocial manner that prefigures the outcome of liberal theorizing. According to communitarians, because individuals come to be within the context of community, and because communities come to be within the context of powerful and far-reaching states, the proper theory of state authority cannot rest on an ontology of atomistic individualism. Instead, it must acknowledge and accommodate the inevitable social embeddedness of individuals and the significance of supra-individual entities, such as religious groups, generational cohorts, culturally integrated regions, and so on. Communitarians are thus more willing than traditional liberals to tolerate the expression and mediation of values, goals, and preferred ways of living in the public sphere through group recognition and discourse. This is not to say that communitarians support an authoritarian or totalitarian conception, in which individuals *must* follow aims or attitudes that are articulated through collective mechanisms. But it is to say that such collectively articulated aims or attitudes are always present in the communitarian's ontology and that they therefore must be reckoned with directly in political life, rather than cordoned off into nominally "private" or depoliticized spheres of interaction. The communitarian approach to public education, for instance, does not limit the government's role to inculcation of a minimal set of "neutral" liberal values, but rather permits public educators actively to support the continuation of group histories and identities.

education and group identity

Liberal theorists have responded to the communitarian critique in various ways.[12] John Rawls, for instance, defended his use of an abstract, unsituated self as a heuristic device for evaluating political norms, rather than as a metaphysical proposition concerning the actual nature of individuals.[13] More ambitiously, Jürgen Habermas relies upon an understanding of the self as situated within a distinctively *liberal* community and argues that "[t]he choice between 'individualist' and 'collectivist' approaches disappears once we approach the fundamental legal concepts with an eye toward the dialectical unity of individuation and socialization processes."[14] Still others contend that, regardless of the merits of the communitarian's characterization of liberalism as unduly individualistic, the implications of communitarian alternatives have yet to be worked out in a way that safely avoids the oppressive potential of collectivism. The force of this last position is almost self-evident; after all, the horror of collectivism gone wrong *is* the lesson of the twentieth century, and liberals of all stripes are right to apply their own principle of precaution against it. Nevertheless, because communitarianism endows supraindividual groups with powers of agency and self-expression to a greater degree than conventional liberalism does, the theory occupies an important, independent locus on the spectrum of agent-relativity. Significantly, even the state itself may be conceived of as a community on this approach, capable of extending itself across time as a distinctive agent with a particular identity and history.[15]

As suggested by the foregoing thumbnail sketches, one obvious and important task that arises from figure 2 is to determine which site on the spectrum of collective agent-relativity reflects the most normatively desirable political theory. For reasons that will be discussed in this chapter, many conclude that the criticisms traditionally lodged against utilitarianism for individuals fail to apply at the level of the state and that, instead, an optimization rubric of the sort underlying welfare economics is especially desirable as a philosophy for government conduct.[16] The metaethical points raised in chapter 1 still remain, however, for some normative justification other than causal optimality is necessary to ground the conclusion that social actors should conform to an optimality standard. At bottom, the various arguments offered in favor of utilitarian approaches to public policy making reduce to statements that social actors and institutions can pursue causal optimality with fewer constraints and unintended side effects than individuals can. Following the ethical agency argument on the individual level, however, it is not enough to say that social actors and institutions should optimize simply because it is the socially optimal thing to do. Not only is this statement sometimes empirically false, in that the deliberate pursuit of optimality is not always the surest route to its achievement, but the statement also is conceptually problematic, in that it is incapable of explaining the embedded assumption that social institutions should do anything—that is, that the decision making of such institutions should be the subject of norms of any sort. The reason for *that* assumption, it must be conceded, is that we implicitly recognize such institutions to be separate and coherently formed agents rather than simply passive instruments of optimization. In the liberal tradition, for instance, we often regard such institutions as bearing a burden of justification to their citizens, to account for the reasons behind public policy choices in a manner that respects the plurality of values held by individuals. On this account, "We the people" never functions merely as a pass-through vehicle for individually inscribed values or interests, since any collective act must be persuasively presented to the citizenry in order to attract legitimacy as a representation of democratic choice. To be sure, during periods of "normal politics" the state may appear to be comprised simply of institutions that govern a mass of disinterested and disconnected private individuals; nevertheless, during constitutional crises and other periods of "higher lawmaking" the popular sovereign reappears on the scene, asserting more directly the collective authorial power that the people always retain.[17]

For some observers, these issues necessarily devolve into fundamental debates concerning the ontological status of the polity. Is it meaningful and appropriate, in other words, to describe the state as a person, an organism, or some other anthropomorphized entity? Are states the kinds of things that can hold goals, espouse beliefs, or generate reasons for acting? Leading legal scholars Eric Posner and Cass Sunstein rather dogmatically assert that nations "do not have mental states and cannot, except metaphorically, act."[18] Such summary dismissals seem to emanate from an assumption that an agent must have a brain in order to have thought or intentionality, a version of the same biophysical reductionism that we saw in the previous chapter. In fact, however, a substantial literature in philosophy, political theory, and sociology addresses precisely the questions of what the state *is* and whether it can be regarded as capable of thought or intentionality. And while some commentators agree with Posner and Sunstein's conclusion, a healthy number and variety do not.[19] A more organicist conception of the state, for instance, featured prominently in the writings of Hegel and Rousseau,[20] and it has recently been given a powerful restatement by Jed Rubenfeld as part of his argument that written constitutionalism helps a democratic people honor its intertemporal commitments in the manner befitting a free agent.[21] The social constructivist theorist Alexander Wendt likewise has advanced a strong set of arguments in favor of the view that states are "purposive actors with a sense of Self."[22] He has even suggested that states may be regarded as organized forms of biological life and as subjects of conscious experience, at least according to plausible interpretations of those concepts that are found in philosophy and the biological and cognitive sciences.

Given the existence of such debate, it may be unwise to assume a posture of ontological essentialism in which states and persons must be defined according to what they are *in fact*. Instead, it may be more profitable simply to examine the social and linguistic practices that work to constitute those entities and to consider what those practices reveal about our political morality. From that vantage point, we see that "states," "humanity," "the market," "global civil society," and a variety of other actors *are* apprehended and discussed as entities with personlike characteristics, for example, awareness, intentionality, and responsibility. Indeed, we might even join Patrick Thaddeus Jackson in his provocative claim that not only are states people, but "*people are states too*, inasmuch as both are social actors—entities in the name of which actions are performed—exercising agency in delimited contexts."[23] Significantly, the welfare economic position is not exempt from this ordinary language argument: Even to speak of a "cost-benefit state" is to speak of an entity about which it makes pragmatic sense to ascribe a method of decision making *and* a responsibility to follow that method. As noted above, proponents of cost-benefit analysis can in fact be seen as closet collectivists who seek to bind citizens together into a welfare-maximization compact. Yet, as the remainder of this chapter argues, the proponents lack a coherent account of how the cost-benefit community can maintain commitment to maximization, given that the cost-benefit state has no higher language available to justify itself or its policies to citizens. Indeed, the staunchly depersonalized state implied by welfare economics gives rise to another provocative claim: On the cost-benefit account, not only are states not people, but *people* may no longer be people, inasmuch as they are made to reside in a community that is bereft of resources for understanding and debating the morality of choice and action.[24] As will be seen, once government officials are reduced to passive agents of a maximization machine—and once citizens are reduced to the input/output function of that machine—the very reason for maximizing in the first place can no longer be apprehended. Thus, even if Posner and Sunstein turn out to be correct to deny the personhood of states as an ontological matter, their truth may come at an unacceptable price.

People no longer are people.

*

As with theories of normative ethics for individuals, it is essential not to confuse the question of how one constructs a conception of collective agency with the separate question of what normative theory should guide the conduct of the agent so conceived. A satisfactory answer to the former question is necessary before the latter even can be entertained, and a peculiar defect of optimific theories such as utilitarianism and welfare economics is that they fail to grant the agent a secure location from which to survey the landscape of ethical possibility. A powerful demonstration of what is at stake in these debates can be seen in Cass Sunstein and Adrian Vermeule's argument that states might be obliged to adopt the death penalty, given some evidence suggesting that capital punishment deters more death in the form of averted criminal homicide than it requires in the form of state-sanctioned killing.[25] With lives seemingly positioned on both sides of the death-penalty debate, thanks to the emergence of this (highly contested[26]) deterrence literature, Sunstein and Vermeule argue that instrumentalist optimization is the only defensible route available to the state. This is because, in their view, "any de-

Death Penalty Debate

ontological injunction against the wrongful infliction of death turns out to be indeterminate on the moral status of capital punishment if the death is necessary to prevent significant numbers of killings."[27] Implicitly, then, the authors see no ethically relevant distinction between a death that the state causes itself, through its appointed agents and according to rules and procedures that it establishes, and a death that occurs at the hands of a private citizen acting in contravention of the state's criminal law. Indeed, they write that with respect to a state, "foregoing any given execution may be *equivalent* to condemning some unidentified people to a premature and violent death."[28] By adopting this equivalency thesis and by assuming the veracity of the empirical literature on the deterrence benefits of capital punishment, Sunstein and Vermeule are led to offer a startling claim: "States that choose life imprisonment, when they might choose capital punishment, are ensuring the deaths of a large number of innocent people."[29] And because no reasonable state should prefer a policy that ensures more deaths than it aims to prevent (let alone more deaths of innocent victims), the authors conclude that the death penalty might well be "morally compelled."[30]

equivalency standard...

Although not framed in terms of welfare economics or cost-benefit analysis, Sunstein and Vermeule's argument trades on the utilitarian philosophy that lies at the heart of those methodologies. It assesses state policies solely according to their predicted consequences for the overall well-being of individuals, attaching no special significance whatsoever to the role of the state as an agent. The authors do try to contend that their argument transcends philosophical disputes between utilitarians and deontologists. However, their contention in this respect depends on an unconvincing assertion that if deontology assumes a moderate form, in which "the numbers" are permitted to matter to some extent, then the approach simply collapses into utilitarianism. What a moderate deontologist would take as ethical *permission* to maximize welfare by doing harm in some exceptional circumstances, Sunstein and Vermeule treat as a "consequentialist override" that *mandates* the doing of harm with the same sort of compulsiveness as their preferred brand of utilitarianism.[31]

This collapsing of moderate deontology into utilitarianism causes the authors to overlook important features not only of capital punishment but also of environmental, health, and safety law. For instance, citing the literature on risk-risk tradeoffs, Sunstein and Vermeule argue that "[i]t is implausible to say that, for moral reasons, social planners should refuse to take account of such tradeoffs." Accordingly, the authors claim that "there is general agreement that whether a particular substance ought to be regulated depends on the overall effect of regulation on human well-being."[32] This phrasing, however, is inexact: Few would contend that social planners should "refuse to take account of such tradeoffs." The thorny issue, the important issue, concerns how best to take their account. Under Sunstein and Vermeule's conception, risk-risk tradeoffs can be accommodated by the state only through a reflexive empirical mandate. They write that "[t]he crucial question is what the facts show in particular domains"; that "[e]verything depends on what the facts turn out to be"; and that "[a]rguments about policies . . . are hostage to what the facts turn out to show in particular domains."[33] Elsewhere, the authors insist that, although "there is a reasonable question about whether capital punishment really does provide net social benefits," it is a question that calls solely for "empirical evidence."[34] Under a less formalized, more agent-relative approach, on the other hand, the resolution of tragic tradeoffs would never be reduced solely to an empirical question. Instead, the deciding agent would always remain cognizant of the unavoidable burden of discretion and responsibility that lends a tragic cast to capital punishment, environmental law, and other areas of regulated violence. Thus, a moderate deontologist might agree with Sunstein and Vermeule that, under some regrettable circumstances, an intentional harm such as capital punishment is justifiable. Nevertheless, the process by which that conclusion will be reached—and the expressive connotations that will flow from the harmful act once it is undertaken—are vastly different than under the utilitarian approach. To nonutilitarians, doing harm always must be done with solemnity.

One can vividly illustrate the differences between these two approaches by returning to Thompson's organ harvesting hypothetical from chapter 1. Suppose that Adrian, a thorough-going utilitarian, were confronted with the decision of whether to sacrifice his single patient in order to save the five who sit desperately in need of organ transplants in his waiting room. Assessing the situation through his empirical lens, he might conclude that, although killing the one undoubtedly is a setback for the world and some danger exists that patients will be less inclined to

[handwritten margin note: Environmental law as regulated violence]

seek medical services if they know that they may be conscripted into the maximization machine, a comprehensive utilitarian evaluation nevertheless requires him in this case to sacrifice the one. Next, suppose that Cass, a moderate deontologist whose ethical theory permits him to cause harm in extraordinary circumstances, were confronted by the same dilemma. He might also determine, after agonizing deliberation, that killing the one in this situation is ethically permissible. Accordingly, if judging only in terms of what outcome is chosen in a given context, one might conclude that the two ethical approaches are indistinguishable. But to examine only outcomes in this manner is to ignore everything that is interesting about the debate between utilitarianism and deontology. Indeed, it is to avoid the debate altogether, for a sea of difference exists between Adrian, whose ethical theory absolves him of responsibility for the death of the one and even deprives him of any sense of discretion and agency in the matter, and Cass, whose ethical theory makes patently plain that matters of life and death are in his hands and requires that all killing, even if justified, be treated with recognition and remorse. The latter approach pays due respect to the importance of agency in underwriting a normative ethical theory, while the former seems to advance a disembodied pursuit of optimization . . . just because.

Sunstein and Vermeule in part recognize this dilemma, but, as noted above, they regard the notion of ethical agency as either unnecessary or incoherent at the collective level. Thus, on their account, public policy making can be made to depend solely on impartial empirical assessment without courting the agency-undermining effects that such an approach threatens on the individual level. Their argument in this regard deserves careful attention, for the desirable degree of agent-relativity at the individual level should not be taken uncritically to supply the desirable degree at the collective level.[35] A variety of important distinctions between the two types of human actors might well lead us to believe that the utilitarian ideal can be better realized at the collective level and therefore should be imposed as an *übernorm* for public policy making. After all, the classic objections to fully impartial utilitarianism—that it disrupts a life filled with projects and meaning and that it precludes special affiliations to family, friends, and compatriots—seem inapplicable to collective institutions that are charged not with crafting an individual identity or with nurturing personal relationships, but instead with serving the overall good of society. Such institutions, one might argue, should be conceived and operated without their role being seen as separate or distinct from the larger causal system within which they operate; collective responsibility instead should extend to the fate of the entire system, unmediated by a "raft of baggage of personal attachments, commitments, principles and prejudices."[36]

Practical considerations also support the promotion of utilitarianism as a public philosophy. Unlike individuals, who could never fulfill a duty of causal optimality with anything other than gross incompleteness and inaccuracy,[37] human collectives are thought to be better able to realize such a duty. Institutional actors can promulgate rules and distribute costs in a broad-sweeping manner that individuals in their private lives cannot replicate, thereby lessening the burden of a duty on the state to aid those who suffer.[38] Also, as Michael Green argues, "[i]nstitutions are better than individuals at collecting and processing information about the distant or indirect consequences of their actions."[39] Indeed, not only can institutional actors countenance a much greater spatial scope of concern than individuals can, but they also can adopt a greater temporal scope of concern, given their legal immortality.[40] Finally, as Sunstein and Vermeule argue, "institutional agents are not hampered by our phenomenal distinction between action and omission" and can accordingly "collect and use information about the consequences of their omissions to a greater extent than individuals can."[41] Put differently, the inadequacies of ethical imagination that seem to afflict individuals, leading them to fail to perceive the manner in which their causal influence implicates the suffering of distant populations, can be remedied at the collective level in a manner that renders the duty of optimality more perfectly dischargeable by states.

For these reasons, a rejection of the act-omission distinction that underlies libertarianism as a public philosophy would compel not only greater use of the death penalty (at least on Sunstein and Vermeule's reading of the empirical evidence) but also a range of progressive state policies not typically associated with schools of ethics and politics that condone capital punishment. As the authors write, "In many situations, ranging from environmental quality to appropriations to highway safety to relief of poverty, our arguments suggest that in light of imaginable empirical findings, government is obliged to provide far more protection than it does now, and it should not be permitted to hide behind unhelpful distinctions between acts and omissions."[42] These multivalent political implications of welfare maximization recall the manner in which utilitarianism on the individual level is associated both with cosmopolitanism and with rampant dehumanization, with Singer's powerful arguments in favor of greater levels of global altruism and with ruthlessly calculative, instrumental use of human life. Such seemingly irreconcilable implications of utilitarianism underscore the importance, again, of distinguishing between two different uses of agent-relative conceptions, such as the act-omission distinction: their use as a source of prescriptive advice regarding how to behave in particular contexts and their more general function in helping to underwrite the sense of normativity that causes agents to want to be moral at all. Utilitarianism often seems to outperform deontology with respect to the former use because it requires affirmative obligations to aid and comfort; with respect to the latter use, however, its destabilizing potential becomes more apparent, as does the enduring appeal of nonoptimific frameworks, such as deontology for individual normative ethics or rights-based liberalism for political organization.[43]

Like many commentators, Sunstein and Vermeule do not distinguish between these two different levels of analysis when they examine the act-omission distinction. For Sunstein and Vermeule, practical considerations concerning the reach and power of the state lead to the strong conclusion that its sphere of responsibility should be thought of as coextensive with the entire causal order. Boldly, the authors argue that the idea of an act-omission distinction or similar agent-centered restriction at the level of the state "may not even be intelligible" and is, at least, of "obscure" moral relevance.[44] Their argument again collapses two conceptually distinct matters. Sunstein and Vermeule are certainly correct to note that "[a] great deal of work has to be done to explain why 'inactive,' but causal, government decisions should not be part of the moral calculus [of public policy making]."[45] However, the authors' construction simply replaces one underdetermined notion ("inactive") with another ("causal"). It is true that institutional actors typically hold a broader scope of causal potential than individual actors. That reality, however, does not render the concepts of agency and causal capacity unintelligible or obscure at the collective level. After all, the same challenge that exists on the individual level—the challenge of pursuing desirable outcomes when the agent's causal potential is both filled with opportunities for act-ing and simultaneously constrained by the omnipresence and power of other causal forces—also exists on the collective level. Even robust insti-tutional actors, such as states, confront a phalanx of forces that lie beyond complete prediction and control, including the unpredictable operations of complex biophysical systems, the independent policy choices of other political entities, and the call of unknowable others, such as future gen-erations and nonhuman life-forms, whose well-being depends critically on the state's actions but whose ability to articulate needs and interests in the manner demanded of them by liberal political theory is insurmount-ably frustrated.

At times, the challenge of enacting a political identity within this context becomes dramatically apparent. In the aftermath of Hurricane Katrina, for instance, one anonymous White House official initially sought to deflect criticism of the federal government's response by argu-ing, "Normal people at home understand that it's not the president who's responsible for this, it's the hurricane."[46] Four days later, with criticism mounting, President George W. Bush embraced the opposite normative extreme, in which the scope of the government's responsibility appeared to be limitless: "[A]s long as any life is in danger, we've got work to do. . . ."[47] As a statement of government responsibility for hurricane pre-vention and disaster relief, the latter quotation seems preferable to the former, which trades on a prescriptive version of the act-omission distinc-tion that should be rejected. What both quotations share, however, is an acknowledgment of the state as an independently significant actor, one for which even an apparent duty of comprehensive lifesaving can only be imposed as a result of reasoning, choice, and responsibility. Thus, al-though arguments against agent-relativity at the level of the state are cor-rectly motivated to the extent that they recognize greater causal potential and therefore greater responsibility for the prevention of suffering by states, they overshoot to the extent that they seek to deny the state separ-ateness or coherence as a distinct human institution.

This overshooting is readily apparent in Sunstein and Vermeule's death-penalty argument, for the authors assume an infallible and omni-potent capacity for prediction and control on the part of the state. At one point, they argue against the notion of an act-omission distinction for governments by claiming that the state always has the option to completely deter murder: "[T]he distinction between authorized and unauthorized private action—for example, private killing—becomes obscure when the government formally forbids private action but chooses a set of policy instruments that do not adequately or fully discourage it."[48] By assuming that the state could "fully" deter murder if it would just adopt the optimal set of policies to do so, Sunstein and Vermeule posit an actor with such all-encompassing causal potential that the act-omission distinction seems to disappear before our very eyes. But all of the work for this disappearance act is done in their undefended and implausible assumption of perfect causal potential. Elsewhere they reinforce their omnipotent state device by citing Philip K. Dick's science fiction story *The Minority Report,* in which a futuristic crime prevention bureau is thought to be able to perfectly prevent murder by using psychic "precogs" to foresee when a killing will occur and, thus, to enable detention of would-be perpetrators before they actually commit any crime.[49]

To most readers, *The Minority Report* represents a dystopian tale in which a panoptic bureaucracy threatens to neuter the human spirit, at least until the supposed determinability of human action and infallibility of governmental prediction are exposed by the story's protagonist as unfounded tools of thought control. This reading seems lost on Sunstein and Vermeule, however, as their own panoptic vision of public policy making does not blink. Faced with a slippery slope of nightmarish scenarios, such as the possibility that randomly murdering an *innocent* citizen might be even more effective at deterring crime than capital punishment, Sunstein and Vermeule remain relentlessly utilitarian: "[E]verything will depend on what the facts show and on the costs and benefits of alternative policies."[50] Confronted by the claim that governments might usefully be personified to the extent that we evaluate the intentionality or purposivity of government policies, the authors respond that such concepts only apply to the individual actors within official positions: "Moralized talk about whether 'governmental' intentions are culpable is metaphorical shorthand for institutional criticism and for moral criticism of particular individuals who happen to occupy government posts."[51] Asked to consider the challenge of satisfying the optimific duty in a complex causal order that contains forces beyond the control of the state, the authors again look inward to individual government officials rather than to the state qua state.[52] On Sunstein and Vermeule's account, then, the system of governmentality has no meaningful agency, intention, or limit; it just *is* an all-encompassing program for the good of optimality . . . and what is particularly good about that cannot be said from within their philosophy.

*

precautionary principle = recognition of self as self?

Unlike the cost-benefit state, the political community that adopts and implements a precautionary approach does so with a recognition of itself as a member of a larger geopolitical and temporal community of communities. On the precautionary account, environmental, health, and safety regulation is not merely an opportunity to maximize an existing set of individual preferences or interests, but rather a moment to consider the regulating body's obligations to its present and future members, to other political communities, and to other species. Such notions of decidedly collective responsibility are well demonstrated by the original German articulation of the precautionary principle, *Vorsorgeprinzip*, which translates literally as the "principle of prior care and worry" and which includes notions of "caring *for* or looking *after,* fretting or worrying *about* and obtaining provisions, or providing *for.*"[53] Through these

German Definition

relational constructs, the precautionary principle offers a subtle but constant reminder that the relevant political community's decisions express a collective identity—an identity that the community must in an important and unavoidable sense *own.*

Thus, contrary to prominent critiques, the precautionary principle does not imply a fallacious belief that the larger causal fabric is benevolent[54] or that human omissions are perfectly innocuous.[55] Nor does it necessitate a return to the mistaken view that a stable balance of nature exists beyond the influence of humans.[56] Nor, finally, does the precautionary principle require abandonment of the attempt to foster specific positive duties at the societal level (that is, institutional duties to act on the opportunity to prevent or alleviate suffering). The precautionary principle does, however, imply the view that human agents, whether individual or collective, bear responsibility in a way that other causal forces do not; accordingly, their decision making should be conducted with a sense of urgency and self-awareness. Denying such a notion in favor of an optimization rubric would risk sliding into what Nagel has described as "a standpoint so removed from the perspective of human life that all we can do is to observe: nothing seems to have value of the kind it appears to have from inside, and all we can see is human desires, human striving—human valuing, as an activity or condition."[57]

Ultimately, the fact of collective identity and responsibility is not one that welfare economic proponents deny. Like the evolutionary psychologists who dismiss moral philosophy's motivational power, yet who themselves offer reasoned proposals for behavioral reform, a conceptual tension exists within the intellectual program of welfare maximization. Even as we try to program our institutions to be "hostage to what the facts turn out to show in particular domains," as Sunstein and Vermeule would have it, our very act of programming concedes the ethical distinctiveness of our institutional creations—and the possibility that they might be programmed according to other visions of societal flourishing. Deep below the push for welfare maximization, therefore, seems to lurk the same conception of collectivity that the methodology's proponents regard as suspect within the precautionary principle. For either approach to have compelling, persuasive, or even recognizable significance as a framework for decision making, it is necessary first to conceive of political communities as distinctive agents that can respond to reasons, articulate goals, and maintain self-awareness regarding the urgency of public policies. A necessary predicate for *that* conception, in turn, is to reject the insistence that collective decision making can be reduced to a ministerial act of aggregation. Try as we might to deny it, the embrace of "Government house utilitarianism"[58] is much more than a practical decision involving the institutional satisfaction of individual interests. It is a choice that reveals something intimate and foundational about our collective identity—something that will be lauded, lamented, or viewed indifferently by foreign nations, future generations, and other onlookers, but that will always be seen as uniquely ours.

*

The next two parts of this book underscore the practical importance of holding on to this conception of collective agency. They demonstrate ways in which even powerful institutional actors, such as states, are limited in their ability to optimize the effects of their causal capacity and are therefore required to develop a fuller conception of their own subjectivity than is permitted by optimization models. Most obviously, every state resides under the influence of natural systems that escape precise probabilistic

understanding, a challenge made all the more difficult by growing awareness that, at least for certain complex, adaptive systems like the atmosphere, improved understanding of their operations may actually lead us
to be *less* confident in our ability to predict and control (chapter 3). States
also are confounded in their optimization efforts by the fact that certain
human values and aims cannot be decomposed into a mere collection of
individual interests, but rather those values and aims in part emerge
from the processes of social and political interaction themselves (chapter
4). Because endogenous collective influence of that nature is inescapable,
the attempt to identify and implement a social welfare function often suc-
ceeds only in avoiding difficult questions about the kinds of preferences
or values that we wish to promote among individuals.

Concerted ethical—and not merely empirical—engagement is also
demanded in relation to the following factors: the choices and actions of
foreign nations that both depend on and impact shared resources and
that, because they emanate from reasoning agents, cannot simply be reduced to another deterministic element of the causal order (chapter 5);
the impact of policy decisions that will be felt in the future by generations whose identity and interests we can scarcely glean using our conventional models of assessment and aggregation (chapter 6); the elusive
utterances of nonhuman life-forms, which also are incapable of expressing their interests in the manner demanded of them by our dominant
political theories (chapter 7); and, most challenging yet, the prospect of
engineered life-forms whose moral and political standing will require
our prior acknowledgment—according to normative considerations we
have scarcely begun to articulate—before they may enter into our aggregation exercises (also chapter 7).

To reiterate, the basic claim of this chapter is that precautionary approaches to environmental, health, and safety regulation display greater
sensitivity to the inevitable situatedness of collective decision making
than do optimization models such as welfare maximization. By reminding the political community, as it stands ready to undertake an action
with potentially serious or irreversible environmental consequences,
that the community's choices express an identity that will belong to it
alone, the precautionary principle helps to sustain appreciation for the
normativity of policy analysis. That is, it helps to remind the community
of why the results of policy evaluation exercises have significance, not

merely as pieces of information about possible courses of action, but as reasons for choosing and acting in particular ways. Interest aggregation models such as cost-benefit analysis cannot support their own prescriptivity in this manner; they depend for their persuasiveness on the existence of a kind of collective reasoning subject that is, nonetheless, denied any foothold in the models' representation of social reality. Like the experimental subject described in chapter 1 whose "abnormally 'utilitarian' judgment" led eventually to his unemotional statement of surprise— "Jeez, I've become a killer"—societies programmed according to the calculative logic of utilitarianism also seem capable of losing themselves, of offering up their agency as a "hostage" to empirical happenstance. Down that road lies a wealth of progressive improvements to overall well-being, to be sure. Further along, however, lies an erosion of social and political connectivity, so much so that one day we may feel mysteriously compelled to murder the guilty—and even to sacrifice the innocent— simply as part of our technique of optimization.

Jeez.

PART TWO THE PERILS OF PREDICTION

CHAPTER THREE

Complexity and Catastrophe

RATHER THAN EMERGING from collective deliberation by a political community—from that community's point of view—policies adopted under optimific approaches such as welfare maximization are said by their proponents to "inevitably and predictably" flow from the calculated effects of state action.[1] As the previous chapter argued, even accepting that adequate knowledge is available to perform this formalized conception of policy making, it is unclear how the results of such an approach can retain authority over time, given that the welfare-maximization framework implicitly denies the distinctiveness of its own audience. That is, rather than appearing within the framework as a responsive—and responsible—subject of ethical reasoning, welfare maximization's political community instead becomes simply part of the propwork of the optimization paradigm, the underlying normativity of which is rendered indefinite, even incomprehensible, to the community thus represented. Besides being analytically unsatisfying, such a conception seems to risk a kind of moral anesthetization, a dulling of appreciation for the pressing questions posed by environmental lawmaking precisely at the moment when self-awareness and sensitivity are greatly in need.

This chapter begins the process of recovering the questions not answered, the questions positively obscured, by the risk-assessment-and-cost-benefit-analysis paradigm. It focuses on the most obvious set of such questions, those concerning whether and how policy makers can arrive at the body of empirical data that is demanded of them in order to pursue optimal environmental outcomes. Despite the steadfast claim of scholars such as Sunstein and Vermeule that "[e]verything depends on what the facts turn out to be," the welfare-maximization approach to policy making is curiously silent about the processes by which the "facts" are discovered; it seems to assume that empirical inputs into policy making simply arrive, having been generated through investments, discoveries, and procedures that reside outside the public's influence yet that somehow reliably produce the data necessary to optimize. Any adequate approach to environmental policy, however, must offer insight into the processes of knowledge generation themselves, addressing such questions as: How will environmental, health, and safety threats be identified over time and across changing circumstances? What share of public and private resources will be devoted to research into the causes and consequences of such threats? Which actors will bear the burden of proof regarding demonstration of the harm or the desirability of proposed activities, technologies, or substances? Questions of that nature demand a theory, not of substantive but of *procedural* rationality, quite a different orientation from the end-state focus of utilitarianism and welfare economics.

procedural rationality

In recent years, the need for this kind of epistemologically sophisticated approach to policy making has become all the more apparent, as our conventional mode of representing natural and social systems has come to appear increasingly simplistic and misleading. As J. B. Ruhl observes, "[t]he prevailing schools of environmental policy have described our problem as a series of linear, one-dimensional decisionmaking systems," an approach that assumes "economic conditions can be translated predictably into economic conclusions that call for prescribed economic measures, [and] environmental conditions can be translated predictably into environmental conclusions that call for environmental measures."[2] If these prevailing schools were correct that natural and social systems are well behaved—so that they follow linear operating rules, map onto known probability distributions, and exhibit stable equilibrium outcomes—then data gaps and other shortcomings of human knowledge would not be deeply problematic: Standard modeling assumptions and related technical conventions could be relied on to transform situations of ignorance and uncertainty into situations of risk (fig. 3),[3]

	Outcomes Well Defined	Outcomes Poorly Defined
Probabilities Well Defined	**Risk** *e.g.*, routine floods, highway safety, known diseases	**Ambiguity** *e.g.*, some climate change scenarios
Probabilities Poorly Defined	**Uncertainty** *e.g.*, floods under climate change, many carcinogens, catastrophic influenza	**Ignorance** *e.g.*, thermohaline circulation shutdown, endocrine disruptors

FIGURE 3. Varieties of Knowledge Conditions

thereby preserving at least the nominal coherence of optimization as a modus operandi for policy making.

Unfortunately, many of the systems under inspection in environmental law and risk regulation resist this description; rather than being simple and static, they are complex and adaptive, exhibiting features such as feedback, nonlinearity, and emergence. Thus, environmental lawmakers and regulators not only must assess and manage threats of an unknown magnitude but they also must do so within the context of numerous overlapping dynamic systems, each of which is characterized by such perplexing features as extreme sensitivity to minor variations in condition, irreducible levels of uncertainty, and "fat tail" probability distributions, in which catastrophic events appear with unexpected severity and regularity.[4] Individual components within such complex systems are interconnected through numerous multidirectional avenues of impact, giving rise to multiplier effects and other self-reinforcing tendencies that render a system's condition at any given time both difficult to predict ex ante and difficult to undo ex post.

Complex adaptive systems should be contrasted with the reductionist focus of the Newtonian tradition in science, which attempts to understand the world by breaking it down into smaller and smaller components for isolated study. A central tenet of the study of complexity is that such a "reductionist methodology will never lead to a fully predictive theory of any complex system."[5] For instance, researchers may be able to identify the dose-response curve that characterizes the acute toxicity effects of a given substance on a given species within the controlled splendor of a laboratory environment, but at the same time they may miss entirely the effects of the substance in the species' broader ecological context, where predators, parasites, and other variables absent from the experiment may interact with the substance to enhance its harmful potential, or where the substance may exhibit systemic or long-term effects, such as endocrine-system disruption or teratogenesis, which evade detection within the experiment's truncated time horizon. In addition to highlighting such synergistic, chronic, or otherwise difficult-to-discern causal pathways, complexity theory also posits that microlevel interactions between forces or agents within a system can give rise to properties that only manifest—that only *emerge*—when the system is viewed at the macro level. Minor, even immeasurable variations in conditions between two otherwise identically situated systems, such as the presence in one system of the proverbial flapping of a butterfly's wings, can give rise to dramatic differences in outcome between the two systems only a few evolutionary steps later. The resulting "chaos" is not randomness per se, but rather "order masquerading as randomness," a state of being that, although deterministic, nevertheless remains irreducibly uncertain.[6]

Complexity theory suggests a far more intractable problem setting than has tended to be recognized within environmental law debates, even by those who critique the economic approach for its Herculean informational assumptions. Because complex adaptive systems contain ineliminable uncertainties that cannot be presumed to be of minor importance, such systems by their nature are likely to present ill-posed problems, that is, problems whose imperviousness to resolution is driven not by deficiencies in our epistemic position, but rather by features inherent to the problems themselves. This, then, is an important driver of environmental law's resistance to formalization: We cannot attain the optimization ideal—and thus we cannot effectively disclaim our ethical agency in favor of institutions that are programmed simply and faithfully to optimize well-being—because many of the systems that variously support and threaten our well-being are themselves resistant to formalization. Riddled throughout any attempt to subject environmental lawmaking to empirical technique will be guile and guesswork, moments of policy underdetermination that demand scrutiny, even by a political body that has otherwise committed its organs to ministerial operations of optimization. Such inevitable moments of discretion underscore the need to preserve a coherent sense of collective responsibility, one in which self-awareness regarding our shared agency has not been dissolved into empirical logic, but rather has been called out for unflinching engagement with the tasks at hand.

*

The dangers of failing to promote such broad engagement are well illustrated by a hurricane-protection planning process that was led by the U.S. Army Corps of Engineers in conjunction with local New Orleans levee boards and other government agencies during the latter portion of the twentieth century.[7] Guarded by a diminishing barrier of coastal wetlands and resting on steadily subsiding ground at the confluence of the Gulf of Mexico, the Mississippi River, Lake Pontchartrain, and Lake Borgne, the city of New Orleans occupies a precarious topography. Thus, the city has long relied on an extensive system of levees and barriers to protect it from river flooding and storm surges. Most significantly, in the wake of devastation caused by Hurricane Betsy in 1965, the U.S. Congress authorized the Lake Pontchartrain and Vicinity Hurricane Protection Project (LPVHPP), a vast levee-improvement project that, with numerous delays and controversies, was designed and constructed for New Orleans by the Army Corps over the ensuing decades.

Like most other Army Corps hurricane-protection efforts since the 1960s, the LPVHPP contained at its heart a technical risk-assessment model known as the "standard project hurricane" (SPH). The SPH was developed by the Army Corps at the request of Congress "to provide generalized hurricane specifications that are consistent geographically and meteorologically for use in planning, evaluating, and establishing hurricane design criteria for hurricane protection works."[8] In addition, all Army Corps hurricane-protection projects since 1936, including the New Orleans levee system, have needed to pass a congressionally mandated cost-benefit test.[9] Thus, delving into the origins of the LPVHPP provides a valuable opportunity—made all the more salient by the system's devastating failure during Hurricane Katrina—to examine the welfare-maximization-policy construct, with its twin tools of risk assessment and cost-benefit analysis. As will be seen, moments of empirical

[margin note: SPH is a technical risk assessment model]

equivocation appear throughout the LPVHPP planning process, thereby underscoring the need for some measure of critical distance to be maintained between a political community and its technologies of policy evaluation.

Initially, an overriding goal of the SPH appears to have been simply a desire to compare hurricane-protection standards for different geographic regions.[10] Over time, however, the SPH acquired greater normative weight, being described in Army Corps documents and other contexts as the most severe storm that the government "reasonably" or "practicably" should guard against when designing hurricane-protection projects. Thus, the SPH came to represent not only a method for comparative assessment of storm risks across localities but also a design standard that carried its own implicit assurance of desirability, providing the relevant hypothetical storm in defense of which levee heights and barrier strengths should be calibrated. The following quotations illustrate this normative reification of the SPH:

- "The SPH is intended as a practicable expression of the maximum degree of protection that should be sought as a general rule in the planning and design of coastal structures for communities where protection of human life and destruction of property is involved."
- "An SPH is one that may be expected from the most severe combination of meteorological conditions that are considered reasonably characteristic of the region."
- "The project has been designed to afford complete protection from the occurrence of the largest probable storm (SPH) that can reasonably be expected in the region. . . . Probability of occurrence of hurricanes having a greater magnitude than the SPH are too remote to warrant practical consideration."
- "The project is designed to protect against the 'standard project hurricane' moving on the most critical track. Only a combination of hydrologic and meteorologic circumstances anomalous to the region could produce higher stages. The probability of such a combination occurring is, for all practical purposes, nil."
- "To identify a level of risk a given area faces, we do engineering and an economic analysis and come to an optimum solution for a level of protection."[11]

The veneer of reasonableness and optimality that came to adorn the SPH design standard was unwarranted. To begin with, the SPH could only be as reliable as the data it was built upon, yet the empirical basis of the SPH was both incomplete and untrustworthy. The original SPH model appeared in 1959 and resulted from research undertaken by the Army Corps in collaboration with the U.S. Weather Bureau. The 1959 researchers began their effort by dividing the East and Gulf Coasts of the U.S. into regional zones, with New Orleans falling into Zone B (fig. 4). They then compiled data for each zone on all tropical storms that had reached hurricane intensity at some point during their lifetimes. Storm measurements were available for the years from 1900 to 1956; however, as the authors of the 1959 report acknowledged, the data were somewhat unreliable, given the imprecision of available measurement and recording technologies. In particular, for much of the data, researchers had to estimate offshore statistics loosely from land-based measurements, given that it was not until later in the twentieth century that meteorologists began using aircraft to fly into the eye of offshore storms.

Even assuming valid measurements, the fifty-seven-year record was quite limited in scope, containing only twenty-two storms in total for Zone B, a record that was insufficient to generate anything like a statistically significant rendering of the distribution of potential storms from a multicentury perspective. Moreover, even with respect to the data that were available, one of the most severe observed storms was not used in the construction of the SPH because, according to a footnote in the 1959 report, many of the key model calculations had already been completed by the time the storm occurred. Nevertheless, from this limited data set, the researchers tabulated the cumulative number of storms that appeared during the observation period at or below various levels of central barometric pressure (fig. 5). Generally speaking, the lower the central pressure, the greater the intensity and surge potential of a storm. This measure was then converted into a one-hundred-year index by stretching the data out proportionally from fifty-six to one hundred years. Finally, the data were plotted on normal-distribution graph paper, with the basic idea that if the observed data appeared to fall into a straight line, then researchers could conclude that hurricane frequency follows a normal distribution and, therefore, could extrapolate to longer return periods by extending the observed trend line in a linear fashion (fig. 6).

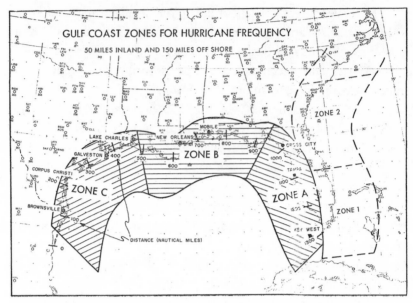

FIGURE 4. National Hurricane Research Program Report No. 33, Geographic Zone Designations

There may be reason to doubt this extrapolation technique. Looking at Zone A, which included Florida and areas east of New Orleans, one observes that, in addition to the much sharper slope of the pressure data, at least one recorded storm lies far outside the normal distribution trend (fig. 7, lower-left-hand corner).

Of course, this is just one storm, and it is difficult to say whether the storm represents a one-hundred-, five-hundred-, or ten-thousand-year event.

But that is precisely the point: With such a small sample, little empirical observation actually supported the researchers' conclusion that storm frequency follows a normal distribution. The extrapolation depended instead on an underlying theoretical conviction that storm behavior fits within the tidy world of classical mathematics. It may well, of course, but it may also reside within what Daniel Farber has called the world of "probabilities behaving badly,"[12] a world in which complex, adaptive systems are characterized not by normal probability distributions but by power-law distributions in which extreme events appear with a surprising regularity.

what level of event does this storm represent?

power-law distributions?

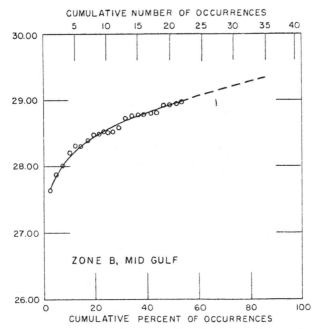

CUMULATIVE NUMBER OF OCCURRENCES

ZONE B, MID GULF

CUMULATIVE PERCENT OF OCCURRENCES

FIGURE 5. Cumulative Number of Storms in Zone B at Various Pressure Levels

Along those lines, consider a few facts from the 2005 Atlantic hurricane season:

- Twenty-seven Atlantic storms were named during 2005, the most on record, shattering the previous record of twenty-one from 1933. For the first time, meteorologists had to reach into the Greek alphabet for additional storm names.
- Fifteen hurricanes were observed, breaking the old record of twelve set in 1969.
- 2005 saw the most Category 5 storms ever recorded in one season in the Atlantic basin (Hurricanes Katrina, Rita, and Wilma).
- Hurricane Wilma became the strongest hurricane on record in the Atlantic basin, as measured by barometric pressure. Three of the six strongest hurricanes on record occurred in 2005.
- Hurricane Katrina made landfall with wind speeds of 125 mph and a minimum central pressure of 27.13 inches, the third lowest on record at landfall.

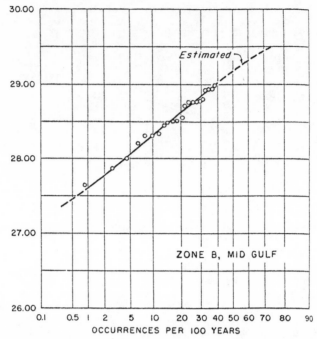

FIGURE 6. Cumulative Number of Storms in Zone B at Various Pressure Levels, Converted to One-Hundred-Year Index

- Hurricane Vince became the first known instance of a tropical cyclone making landfall in Spain.
- Hurricane Delta became only the sixth hurricane on record in December since 1851.
- Tropical cyclone Zeta became the longest-lived tropical cyclone ever recorded in January.[13]

Of course, these facts are merely suggestive; it will be decades, if not centuries, before we can say with confidence whether assumptions of predictable, linear behavior are appropriate for the task of estimating storm behavior in the Gulf of Mexico (or, for that matter, whether the impact of climate change will force us to re-evaluate entirely our understanding of the predictive qualities of the past).[14] Even putting aside such problems of fundamental model uncertainty, however, one still faces the basic policy decision of how conservative to be in setting the benchmark for storm protection. As noted above, the 1959 researchers focused on

Question still is how conservative we should be in setting these estimates?

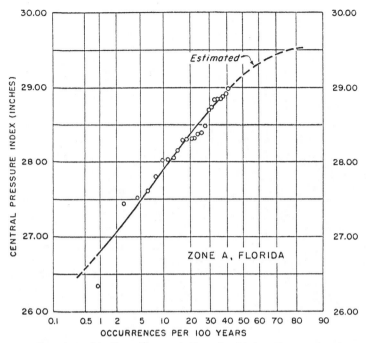

FIGURE 7. Cumulative Number of Storms in Zone A at Various Pressure Levels, Converted to One-Hundred-Year Index

central barometric pressure as their main storm-estimation characteristic; they then constructed a table reflecting the lowest pressure that one would expect to find at various locations with an annual probability of 1 percent. In other words, they chose the one-hundred-year low for central pressure as the primary basis for the SPH standard. Figure 8 shows the resulting pressure values at various geographic locations throughout the Gulf. For New Orleans, the one-hundred-year pressure estimate was 27.60 inches, much less severe than the 27.13 inches recorded as Katrina made landfall in 2005.

Although the SPH model used a one-hundred-year return period for central pressure, it bears emphasizing that the resulting model hurricane did not, strictly speaking, represent a one-hundred-year storm. Instead, the pressure statistic was interpolated with other characteristics, such as radius, wind speed, forward speed, and direction, to generate "the most severe conditions . . . that are within the parameters of the SPH indices . . . for [a particular] location."[15] This procedure explains

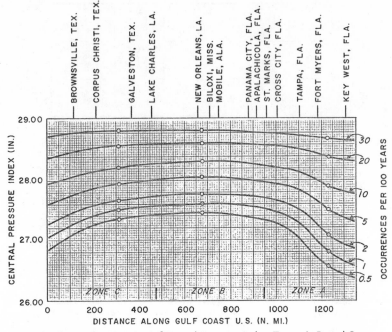

FIGURE 8. Geographic Variation of Central Pressure Index, Zones A, B, and C

why the New Orleans levee system was frequently described after Hurricane Katrina as having been designed to guard against "something like" a two-hundred-to-three-hundred-year storm.[16] In actuality, as the National Weather Service stated in a 1972 technical memorandum, "the standard project hurricane has no frequency assigned to it."[17] Similarly, because development of the SPH model preceded the now familiar Saffir-Simpson Hurricane Scale, descriptions of the LPVHPP using that scale in the wake of Katrina also were ambiguous: Depending on whether one is referring to pressure, radius, wind speed, or some other critical storm characteristic, the SPH can vary from a Category 2 to a Category 4 storm on the Saffir-Simpson Scale, although many post-Katrina commentators described the SPH as "roughly equivalent" to a fast-moving Category 3 storm.[18]

Irrespective of these complications, the question remains as to why the 1959 researchers anchored on a return period of one hundred years when calculating the primary estimation characteristic, rather than, say, five hundred or ten thousand years. As noted in a 1972 revision to the

SPH, the decision to hinge the model hurricane on a one-hundred-year return period for central pressure was essentially an "arbitrary" one, at least when considered from a scientific or technical perspective.[19] This is not to say that the original analysts were unjustified in choosing a one-hundred-year return period or that some other period was obviously more appropriate. It is simply to say that the question did not admit of a purely technical answer. Relevant nontechnical considerations are hinted at within contemporaneous descriptions of the SPH model, where commentators described the model as being used to project the worst storm for which it was "economically []justified" to guard against.[20] In fact, some Army Corps economists at the time believed that the SPH was too cautious and that a less-severe storm should be used as the benchmark for disaster planning and prevention.[21] These economic considerations, however, were not transparent in the model itself, despite its claim to deliver an empirically derived "optimum" level of protection.

This murky blending of science and policy continued in an extensive 1979 overhaul of the SPH. In this report, the SPH was changed so that the critical pressure parameter was derived not from the lowest expected pressure over a one-hundred-year return period, but rather from the average of the seven lowest storms actually observed for a particular geographic area (fig. 9).

This procedure may seem to be an improvement over the essentially arbitrary selection of a one-hundred-year low, but it raises similar questions: Why not take the lowest five storms, instead of the lowest seven? Why not take the single lowest storm? Indeed, why not take the single lowest storm with an added margin of safety? The researchers actually did something quite the opposite, which is that they *excluded* from their seven-storm average the two most intense observed storms: Hurricane Camille from 1969 and the Labor Day Hurricane that struck the Florida Keys in 1935 (fig. 9, points 69 and 35). The reasons provided for these exclusions are somewhat obscure in the report. The researchers wrote, "Our decision was based on the idea that these two hurricanes contained extremely low [pressure values] resulting in sustained wind speeds that were not reasonably characteristic of the northern gulf coast and the Florida Keys."[22] This explanation is unsatisfactory: Excluding outlier data may be standard procedure for many statistical tasks, but the practice seems singularly inappropriate for disaster planning, where the

[handwritten margin note: Excluding outlier data is a bad idea for disaster planning]

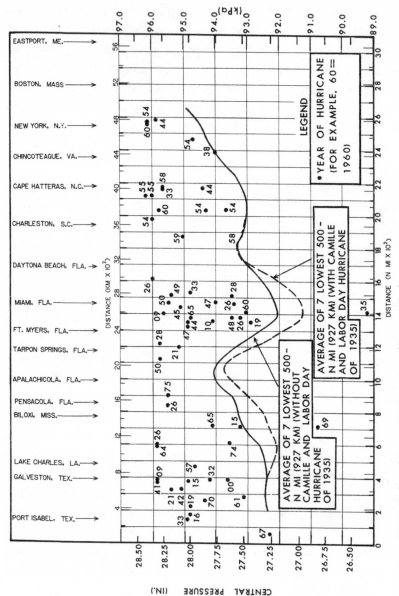

FIGURE 9. NOAA Technical Report NWS 23, Central Pressure Index

extreme tails of a distribution should be precisely the areas of most interest and concern. To underscore this point, consider the fact that the two storms excluded by the 1979 researchers—Hurricane Camille and the Labor Day storm—were the only two Atlantic storms on record with a lower central pressure at landfall than Hurricane Katrina at the time of the New Orleans disaster.

The subjective nature of the SPH construction is implicitly acknowledged elsewhere in the 1979 report, where the analysts recommended use of a more precautionary storm model—dubbed the "probable maximum hurricane" (PMH)—for disaster planning that occurs "in locations where high winds, waves and storm surge could pose a threat to the public health and safety from a hurricane-induced accident at a nuclear power plant."[23] The availability of this PMH model prompts an obvious but unanswered question: Why not use this higher standard of protection for projects that do not involve nuclear power plants but that still concern the safeguarding of human life? More than eighteen hundred individuals died in New Orleans and elsewhere as a result of Katrina's effects, not to mention the thousands whose lives were irreparably damaged. Where were the impact estimations supporting the conclusion that only nuclear power plants merit the higher degree of protection afforded by the PMH? They do not exist because a sub-rosa cost-benefit analysis deemed them unnecessary: As one observer noted, "[t]he design of structures to provide protection against the probable maximum hurricane would, in most locations, be economically unjustified."[24] However, because the SPH purported to provide an objectively determined level of storm protection, such economic considerations weighing in favor of the lower standard of protection never surfaced for scrutiny. Policy makers and citizens were deterred from asking directly whether an extreme event is worth guarding against because the very possibility of the event's occurrence was denied—arbitrarily—by the SPH risk assessment.

The new conservative approach of taking seven of the lowest storms on record did result in a further revision downward, to 27.30 inches, of the New Orleans SPH central-pressure measure. Nevertheless, Katrina's even lower pressure of 27.13 inches at landfall clearly was foreseeable to the designers of the SPH model, as a comparison of the SPH and the PMH estimates for New Orleans demonstrates (fig. 10).

[handwritten margin note: Seven lowest storms approach still does not work]

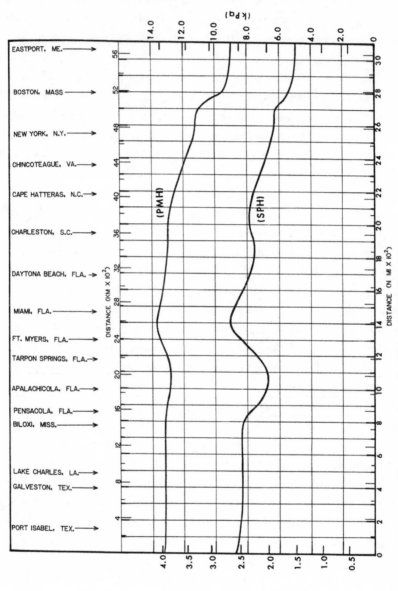

FIGURE 10. Comparison of Pressure Decreases for SPH and PMH Models

The various normative judgments buried within the SPH model construction may have been driven by a concern that the cost-benefit constraint facing the Army Corps' projects would not have justified higher levels of storm protection. No doubt such concerns were themselves driven by the Army Corps' failure at the time to monetize the human health and environmental benefits of its hurricane protection projects.[25] They also likely were affected by the Army Corps' standard use of a 3.25 percent discount rate to translate the anticipated costs and benefits of projects into present-value terms,[26] a procedure that chapter 6 shows to embody a clumsy and inadequate way of addressing questions of intergenerational equity, particularly in the face of long-range planning of the sort implicated by disaster policy. The excluding of saved human lives from cost-benefit calculation and the discounting of future benefits may also have contributed to the Army Corps' tendency to liberally include prospects for private development as part of its flood-control and hurricane-protection projects. Because the Army Corps did not include saved human lives (or, for that matter, ecological resources) in its cost-benefit analyses, the bulk of the identified benefits from hurricane protection tended to come from the safeguarding of real and personal property.[27] Thus, in order to generate a higher regulatory "budget" for project planning purposes, the Army Corps designed projects with easily identifiable and monetizable property-protection benefits, even if that meant earmarking for future development wetlands that might otherwise have remained in their natural, storm-surge-dampening state.[28] Indeed, a key aspect of local opposition to the LPVHPP in New Orleans had centered on the question of whether the corps' design went beyond protecting existing and anticipated land developments to actively promoting new development that would not have occurred but for the corps' activities. As one analyst noted, "[a]n extraordinary 79% [of the net benefits from the LPVHPP] were to come from new development that would now be feasible with the added protection provided by the improved levee system."[29]

Even if the corps had included human health and environmental values within its cost-benefit calculations, the agency still would have faced theoretically and normatively difficult questions regarding how to monetize those values and how to account for their intertemporal

[handwritten margin note: focus on real property changed the nature of the budgeting process]

distribution, as chapters 4 and 6 demonstrate. What the New Orleans hurricane planning process more narrowly shows, however, is that cost-benefit analysis in practice is capable of leading to the very kinds of political and analytical distortions that the procedure is designed to guard against. The failure to account adequately for the lifesaving purposes of hurricane protection led the corps not only to understate the justification for hurricane protection but also to promote private land-development schemes that may well have been counterproductive from the perspective of guarding against storm surges. Welfare economists promote their approach to policy making in part because they believe that citizens and regulators fail to appreciate the unintended consequences of environmental, health, and safety regulation. Yet there is no reason to suppose that their preferred brand of policy analysis is immune from the problem of unintended consequences. Indeed, in this case, cost-benefit analysis seems to have led the Army Corps experts and officials to fall prey to a familiar danger: Because they could not measure what is important, they made important what they could measure.

made important what they could measure

＊

Throughout these decades of adjustment to the SPH model, the actual design and construction of the LPVHPP carried on with similar fitfulness. Bogged down by intergovernmental squabbling, citizen-group opposition, cost overruns, and a host of other problems, the project took far longer to implement than anticipated at the time of its initial authorization by Congress. Nevertheless, by the time Katrina struck in 2005, the LPVHPP was substantially completed; accordingly, in the immediate aftermath of Katrina, many commentators assumed that the storm had simply overwhelmed the SPH design standard, as experts for years had been warning would eventually happen. Some subsequent engineering studies suggested that the primary culprit was not a failure of design but of implementation, such that it is possible Katrina would not have overwhelmed the New Orleans levees had they been constructed and maintained properly.[30] In contrast, an Interagency Performance Evaluation Task Force convened by the Army Corps concluded that Katrina had generated water levels in excess of design criteria throughout most of the storm-protection system and that this overtopping had created significant scour and erosion forces behind the levee walls that enabled their collapse.[31] Irrespective of what ultimately caused the levees to collapse, looking critically at the LPVHPP design process for evidence of short-comings in our thinking about long-term catastrophic threats remains a valuable exercise. The many pre-Katrina warnings that seemed so prophetic in the storm's immediate aftermath remain urgently relevant today, both to the post-Katrina reconstruction process and to the challenge of thinking more generally about natural disasters and other environmental harms in a time of potentially abrupt and dramatic climate change. Most obviously, as the Interagency Performance Evaluation Task Force concluded, the SPH methodology "is outdated and should no longer be used."[32] Rather than the implicit cost-benefit optimization goal of the SPH model, the task force urged that policy makers prioritize a goal of resilience, understood in terms of "the ability to withstand, without catastrophic failure, forces and conditions beyond those intended or assumed in the design." Such a precautionary attitude, the task force noted, was simply "not an element in the New Orleans [Hurricane Protection System] design."[33]

Marshalling support for current public investment in long-term environmental and natural-disaster-prevention projects is a political challenge of the highest order. Given this difficulty, one advantage of the conventional return-period approach to describing flood- and storm-protection projects is its ready cognitive accessibility to nonexpert audiences.[34] Unlike the relative opacity of the SPH model, protection standards expressed in frequency terms comport well with lay individuals' preferred mode of comprehending and evaluating probabilistic information; accordingly, the standards can enable citizens to hold accountable government officials and experts who might otherwise feel immense pressure to contain the costs of protection projects.[35] For instance, when the Netherlands suffered a devastating storm in 1953 that killed two thousand people, the nation embarked on a multibillion-dollar, thirty-year plan to protect the country against the worst storm that could be expected for ten thousand years, a collective project that has become nothing less than a part of the Dutch "national identity."[36] Similarly, in 1927, when a massive Mississippi River flood killed several hundred individuals, displaced more than five hundred thousand, and destroyed property worth some $3 billion, Congress and the Army Corps developed an especially robust flood-control system that was designed to withstand an eight-hundred-year flood, a standard of protection that was some five hundred to six hundred years more forward-looking than the standard adopted for the LPVHPP later in the century.

The wisdom of such protection standards depends in part on engineering and economic factors to be estimated by experts, but also on ethical considerations regarding risk, uncertainty, and intergenerational obligation to be evaluated by citizens. The systems that drive the incidence and severity of environmental disasters are characterized by enormous complexity and uncertainty, whether they take the form of natural systems that give rise to extreme weather and geological events or of social systems that determine in part how deadly and costly the consequences of such events will be. Accordingly, what always will be required in long-term environmental and natural-disaster planning is collective judgment regarding the degree of commitment that citizens desire to express, both to their fellow citizens within the present generation and to those unborn individuals who comprise the generations to come. Because such long-term prevention and mitigation projects have highly uncertain payoffs, their normative desirability is almost never capable of empirical verification, especially in the eyes of the generation that must first embark on a ten-thousand-year intergenerational compact. Instead, that first generation must act on the basis of highly incomplete and imperfect information; its precautionary action must be motivated at least in part by the extra-empirical concerns that give such policy-making contexts normative traction. Rather than highlight such concerns for public scrutiny and deliberation, the SPH instead buried them within a confidently expressed, but ultimately illusory, assurance of "reasonableness" and "optimality."

extra-empirical concerns should be motivators

*

As the New Orleans levee planning history demonstrates, adherence to the synoptic paradigm of risk assessment and cost-benefit analysis requires decision makers to adopt techniques and strategies for rendering policy spaces quantitatively manipulable, even when the empirical basis for doing so is quite thin. By no means are the Army Corps planners alone in such efforts to deny the empirical underdetermination of environmental policy: A recent demonstration of this danger can be seen in certain quasi-scientific attempts by the Food and Drug Administration (FDA) to control the scope of the scientific risk-assessment process for food, drugs, and cosmetics by adopting a presumption that novel technological processes themselves are unworthy of heightened scrutiny. Such an assumption, for instance, seemed to underlie the agency's early determination that nanomaterials in consumer products such as sunscreens do not require an additional risk assessment if their macroscale counterparts have been previously evaluated for health and safety risks. The assumption also may be located in the "substantial equivalence doctrine," which the FDA has used in the genetically modified agriculture context, and the "compositional analysis method," which it has used in the case of cloned livestock for human consumption. In all three cases, the FDA has made its risk-assessment burden lighter by positing that novel scientific processes (in these cases, nanoengineering, genetic modification, and cloning) are not in themselves cause for regulatory scrutiny or distinction, but rather only become relevant if they lead to demonstrated differences in the physical or compositional characteristics of end products as compared to conventional counterparts.[37]

The flaws with such an assumption are many. For present purposes, the most significant shortcoming is the assumption's implicit view that "what we don't know won't hurt us." In the FDA's approach, situations of deep uncertainty regarding the potential impacts of novel technological processes are treated as unworthy of regulatory attention unless some specific material basis for concern is identified, an approach that reflects what Wendy Wagner has called the "unprecautionary principle."[38] To be sure, the resulting permissive approach comports well with the tendency of liberal market democracies to permit private action unless and until a public justification has been demonstrated. The problem, however, is that this predisposition has been presented in a scientific vernacular, as an assumption about the empirical tendencies of nascent technologies, rather than as what it properly is—a preference for distributing the burden of uncertainty in a particular way, according to political values. In this context, the approach happens to be debatable not only on extra-empirical grounds but also on scientific grounds. The FDA's assumption of nano- and macroscale equivalence, for instance, is dubious: Scientists believe that nanoparticles are potentially revolutionary as industrial inputs precisely because they display marked differences in chemical and physical behavior as compared to their macroscale equivalents.[39]

Proponents of the risk-assessment-and-cost-benefit-analysis paradigm have adopted more nuanced ways of dealing with incomplete risk-assessment information. Analysts sometimes contend, for instance, that the proper response to situations of uncertainty is not to abandon the quest for optimization, as the precautionary principle appears to require, but instead to estimate and incorporate the costs and benefits of uncertainty directly into the optimization model. Although preferable to the FDA's blunt refusal to acknowledge uncertainty, this procedure still suffers from a basic limitation, in that the problem of uncertainty will recur unto itself: Without knowing the expected value of future knowledge (which depends on the same unknown probabilities and outcomes that render the situation imperfectly characterized for purposes of risk assessment), the analyst cannot identify the point at which broadened regulatory inspection itself is no longer cost justified. Unwilling to concede the blunt reality of indeterminacy, the analyst instead teeters on the edge of an infinite regress. In the context of complex adaptive systems, this problem is especially acute because the analyst cannot rely on a constant trend of diminishing returns from knowledge acquisition, given the possibility that minor perturbations in one period may give rise to dramatic effects many periods hence.

Introducing genetically modified organisms or nanomaterials widely into field environments raises similar concerns due to the practically irreversible nature of such actions. To address such irreversibilities, proponents of cost-benefit analysis typically argue that the expectation calculus should be expanded to include whatever "option value" would be lost by engaging in an action that poses uncertain and potentially irreversible effects.[40] For instance, one of the earliest and most significant papers in the environmental economics literature began by observing, "[I]f we are uncertain about the payoff to investment in development, we should err on the side of underinvestment, rather than overinvestment, since development is irreversible."[41] Proponents of the precautionary approach would, of course, wholeheartedly agree. They would not agree, however, that the option value of this precaution should simply be priced and incorporated into the optimization calculus so that cost-benefit analysis can continue in "the usual way."[42] To precautionary principle adherents, such an exercise invites exclusionary, technocratic decision making in the face of grave, uncertain collective choices, precisely the context that they believe instead requires inclusiveness, transparency, and candid acknowledgment that ethical choices are being undertaken.

Also susceptible to this democratic critique is Monte Carlo analysis, which is an increasingly prominent method for addressing empirical uncertainties within integrated risk-assessment and welfare-projection contexts. Essentially, Monte Carlo computer simulations evaluate the effects of policy proposals under thousands of possible states of the world, such that, even in the face of uncertainty, analysts may be able to generate hypothetical distributions of unknown probabilities and locate policy prescriptions that predominate over a wide range of potential conditions. Undoubtedly, Monte Carlo analysis is an improvement over previous approaches to the problem of uncertainty, such as the Principle of Insufficient Reason, which simply assumes arbitrarily that two outcomes are equally likely in the absence of evidence to the contrary.[43] Still, even Monte Carlo techniques depend at their base on the specification of certain assumptions about the theoretical nature of unknown probabilities. In keeping with the classical scientific tradition, analysts often specify normal probability distributions in such situations. When applied to systems that behave instead according to the laws of complexity, such assumptions can lead to dramatically erroneous policy advice.[44]

Even more susceptible to this line of critique is Delphi analysis, which consists essentially of gathering subjective assessments of unknown risks from a survey of experts in relevant fields. Through this method, analysts hope to assign a Bayesian prior subjective belief that, in turn, will afford some nonarbitrary basis for taking a "first stab" at calculating expected outcomes.[45] Along these lines, proponents of formalized policy analysis sometimes even deny that there is such a thing as uncertainty, apparently taking the view that if Bayesian, rather than frequentist, probability theory is adopted, *some* number always will be available to the cost-benefit optimizer.[46] The question immediately raised, however, is *Whose* subjective probability assessments will form the basis of the Bayesian exercise? Without devoting careful attention to concerns of openness and participatory legitimacy, policy makers risk obscuring essentially normative judgments through an exercise that, when properly understood, may merit only a weak case for deference to technical expertise.

This democratic critique of data-filling exercises has particular purchase in the case of catastrophic threats. Scientists today largely agree that humanity faces a real but nonquantifiable threat of experiencing one or more of several catastrophic climate-change scenarios in coming centuries.[47] As summarized by the economist Martin Weitzman, the list of such potential catastrophes includes "sudden collapse of the Greenland and West Antarctica ice sheets, weakening or even reversal of thermohaline circulations that might radically affect such things as the Gulf Stream and European climate, [and] runaway climate-sensitivity amplification of global warming due to positive-reinforcing multiplier feedbacks (including, but not limited to, loss of polar albedo, weakened carbon sinks, and rapid releases of methane from the thawing of arctic permafrost)."[48] In addition to these tipping-point scenarios, Weitzman notes several other potential climate-change effects that also strain the capabilities of conventional predictive models, including "sea-level dynamics, drowned coastlines of unknown magnitude, very different and possibly extreme weather patterns including droughts and floods, ecosystem destruction, mass species extinctions, big changes in worldwide precipitation patterns and distribution of fresh water, tropical-crop failures, large-scale migrations of human populations, humidity-nourished contagious diseases—and the list goes on and on."[49] Each of these events would be accompanied by ecological and socioeconomic effects of enormous impact. Given the complexity of the systems involved, however, the likelihood of their occurrence is difficult, if not impossible, to estimate. Instead, we can only say that a realistic threat exists and that our understanding of the threat can be expected to develop in unpredictable ways.

To grapple with such catastrophic prospects, economist William Nordhaus, in his influential cost-benefit assessment of climate change, calculated an estimate of the risks of catastrophic events based on "a survey of experts," many of whom were economists and other social scientists.[50] In the survey, respondents were asked the following question: "Some people are concerned about a low-probability, high consequence output of climate change. Assume by 'high consequence' we mean a 25 percent loss in global income indefinitely, which is approximately the loss in output during the Great Depression. . . . What is the probability of such a high consequence outcome . . . if the warming is 3 degrees C in 2090 . . . ?"[51] Nordhaus then calculated the global "WTP [willingness-to-pay] to avoid catastrophic risk"[52] by multiplying the surveyed experts' probability estimates times the dollar equivalent of a global catastrophe

[margin note: Nordhaus' "survey of experts"]

(made tractable by the fact that Nordhaus had defined catastrophe as a 25 percent loss of global income) and adjusting the result upward to reflect society's general aversion to risk. In this manner, Nordhaus concluded that the world was only willing to sacrifice 1 percent of global GDP in order to avoid the consequences of a catastrophic climate-change event.

This procedure represents a particularly clear illustration of the seemingly irresistible tendency among policy analysts to treat radical uncertainty regarding a potentially catastrophic or irreversible outcome as being equivalent to a low-probability, high-consequence event.[53] Indeed, Nordhaus's survey questionnaire commences with just this equation, despite its lack of logical justification. Just because a threat is uncertain does not mean that it is of low probability. Moreover, the use of a weighted average of the range of possible outcomes works to smooth out much of the tragic texture of decision making in the face of catastrophe. Unlike repeat-play monetary gambles, for which probabilistically determined outcomes provide an invaluable source of information, the expectation calculus seems to provide a poor decision guide for uncertain and potentially catastrophic events. Put bluntly, either the great conveyor-belt system of the world's oceans will shut down, plunging Northern Europe into an awesomely destructive ice age, or it will not. We do not know what the precise probabilities involved are, but given the nature of discontinuity and tipping points, we do know that the expected utility outcome—the weighted average of these extremes—will *not* occur.

To see the inadequacy of this approach, suppose that one of the generic "low probability, high consequence" scenarios described by Nordhaus actually eventuates, in which case human suffering, death, and other adverse consequences deemed equivalent to "a 25 percent loss in global income indefinitely" will befall future generations. Will members of those future generations regard the decision-making process that led us to pose this threat as having been adequately protective of their interests? Is it sufficient that some number of experts—including social scientists who do not actually study atmospheric, oceanic, and terrestrial systems—will have assigned a purely subjective probability figure to currently inestimable but scientifically plausible worst-case scenarios? Is it sufficient that the global willingness-to-pay to avoid such scenarios will have been calculated in a manner that avoids asking any human *directly* how much they would be willing to sacrifice in order to avoid imposing a risk of devastating magnitude on future generations? What are the reasons for assuming that direct human responses to such a question would be less reliable than a method of decision making, such as cost-benefit analysis, that uses a stylized definition of catastrophe and studiously avoids addressing matters of right and responsibility? What are the reasons for treating catastrophic losses indistinguishably from an agglomeration of trivial ones, albeit with a slight upward adjustment for some decontextualized notion of risk aversion? What justifies the assumption that catastrophic events can be smoothed into continuous functions within policy models, rather than remain in the form that they will actually take, as plunges, spikes, flips, and ruptures?

*

In light of the perplexing policy wrinkles caused by irreversible or catastrophic threats, one might be tempted to carve those threats out for special treatment, leaving the risk-assessment-and-cost-benefit-analysis paradigm to serve as the predominant method for evaluating more routine environmental, health, and safety decision making. The teachings of complexity theory, however, suggest that much of our understanding of "routine" risk regulation is misguided. The problem of irreversibility, for instance, should not be seen as restricted to one-shot disaster scenarios. Rather, given the nature of path dependence, some degree of irreversibility should be expected to characterize all decision nodes within complex adaptive systems. Indeed, if the teachings of complexity theory are sound, then environmental, health, and safety dilemmas will, almost by definition, present problems that contain irreversibility, ineliminable uncertainty, and other computationally intractable features. In such contexts, the deliberate attempt to optimize may not represent simply an imperfect but useful aid to decision making, as defenders of the risk-assessment-and-cost-benefit-analysis paradigm often assert. Rather, it may represent a solution concept that is fundamentally mismatched to the problem tasks at hand. We cannot confidently expect that the errors of risk assessment and cost-benefit analysis will cluster around an "optimal" result; indeed, for ill-posed problems—that is, problems whose imperviousness to resolution is not driven by lack of computational capacity but by features inherent to the problems themselves—the very notion of an optimum eludes meaningful description. Accordingly, we should anticipate that the errors of formalized policy analysis may deviate substantially and unpredictably from decision paths that are easily recognized as desirable, if not necessarily optimal, through less-technical decision procedures.

This argument should be distinguished from the long-standing complaint of scholars that risk assessment and cost-benefit analysis can lead to a situation of "paralysis by analysis."[54] This traditional objection—that the enormous informational demands of risk assessment and cost-benefit analysis bog down the policy-making process to such a degree that, when viewed in light of administrative time and resource constraints, the procedures are pragmatically irrational—continues to have considerable force. A second, more-fundamental objection, however, is that the subject matter of environmental law often takes the form of problems that are ill-posed. It is not enough, in such contexts, to address uncertainty, irreversibility, and catastrophe through mechanical adjustments such as option values, risk aversion premiums, or what Cass Sunstein recently has referred to as a human "extermination premium."[55] As seen, these devices often reflect the professional predilections of policy analysts more than they do considered normative judgment. Likewise, it is not enough to present ranges of costs and benefits rather than point estimates, as proponents of the economic approach sometimes advocate in order to respond to the problem of uncertainty.[56] The implications of complexity theory must be accommodated in the very scientific models that are used to generate ranges of costs and benefits. When done properly, such an accommodation often will mean that scientists are only able to offer a series of qualitatively described scenarios that might flow from policy choices—without probability estimates assigned to them—rather than quantitatively depicted but ultimately unhelpful cost-benefit ranges.[57]

In short, for many environmental problems, maintaining adherence to the optimization-policy framework may be deeply unwise. Particularly in the face of complex adaptive systems, which contain feedback mechanisms that can magnify impressively the consequences of even the slightest regulatory misstep, decision makers should abandon the pretense of actually attempting to locate and pursue an "optimal" outcome. Instead, they should present for collective consideration an unalloyed depiction of what is, and is not, known about the possible consequences of human action, so that the political community can consider directly whether it wants to entertain catastrophic or irreversible environmental harm as part of its unique legacy. Within the gap between the aspiration and the actuality of human prediction and control lies an essential motivational truth: It is precisely the inability to optimize our causal agency that helps to sustain our concern for its careful exercise, that helps to reaffirm the sense that human choices and actions matter as central elements of our identity and not merely as heteronomous moments dictated by "what the facts turn out to be." Importantly, so long as the political community refuses to equate its normative identity with the empirical landscape, then the community will remain capable of influencing that landscape's contour and definition. It will be able to deliberately affect which portions of the landscape become visible and, therefore, which portions come to populate our understanding of "what the facts turn out to show in particular domains."

Put succinctly, action cannot be driven solely by consequences, because our knowledge of consequences is itself a function of our actions. This once was understood by environmental law, but lately it has been forgotten.

CHAPTER FOUR

Interests and Emergence

THE U.S. ARMY CORPS' PLANNING efforts in New Orleans represent a particularly crude version of risk assessment and cost-benefit
analysis. The proper response to such failings therefore might not be to
abandon the determination of policy making through quantitative specification but rather to get more rigorous about it, to deploy the latest and
most sophisticated risk-assessment models and valuation techniques, so
that the results of risk assessment and cost-benefit analysis come to better comport with the theoretical ideal of comprehensive, impartial utilitarian evaluation. As this chapter shows, however, the methodological
problems of cost-benefit analysis are not restricted to the version practiced by the Army Corps. One of the most frequently claimed virtues of
cost-benefit analysis is its ability to synthesize vast amounts of empirical
information regarding policy consequences into a single analytical framework. As the mathematician Kurt Gödel famously demonstrated, however, no formal system of this kind can be both consistent and complete;
instead, there will be truths that are expressible within the language of
the formal system but that are only comprehensible from the perspective of a superior system of meaning, one that, in essence, surrounds the
axiomatic subsystem.[1] Because consistency is taken to capture the essence of formal rationality, practitioners of cost-benefit analysis typically
respond to Gödel's challenge by sacrificing completeness; that is, in

trade-off between completeness and consistency

order to maintain the consistency of cost-benefit-analysis outputs, they treat certain decision criteria as externally given, rather than as subjects of direct policy inspection.[2] Accordingly, despite the widespread impression that cost-benefit analysis encompasses a broader range of concerns than more traditional approaches, it is *only* cost-benefit analysis that must, by its very nature, ignore at least some parameters of a given decision-making context. Practitioners of cost-benefit analysis generally attempt to minimize this complication by exogenizing elements that are thought to be of little practical import or that are believed to be adequately addressed by other institutional mechanisms, such as the tax-and-transfer system. The problem with this otherwise sensible strategy is that, increasingly, cost-benefit analysis is being offered for use in choice settings where the variables being exogenized are of deep and unmistakable significance to the very decision under inspection and where the potential role of alternative governance mechanisms is being displaced by the cost-benefit exercise itself.

[handwritten margin note: Cost-benefit is displacing alternative methods of governance?]

The formal analytical nature of the cost-benefit framework gives rise to an additional, more subtle complication. By virtue of its structure, the framework not only must be substantively incomplete—in the sense that it must exogenize some important determinants of welfare as part of the construction and maintenance of its optimization logic—but it also must be capable of denying its own incompleteness. That is, the framework's axiomatic language of interest aggregation and optimization must—again by its very nature—represent itself as having taken account of all pertinent effects of a policy proposal and as having identified an outcome that is singularly optimal. Thus, no matter how sincerely proponents urge that cost-benefit-analysis results represent merely "one input" into the overall policy-making process, the language of welfare economics undermines these reassurances. Because of what might be called [welfare economics' monolinguism,] critics must overcome an unwarranted burden of proof whenever they seek to emphasize the significance of missing interests, to propose the inclusion of missing interest holders, or to challenge the manner in which uncertain outcomes or debatable valuations have been addressed. Invariably, the welfare calculus derides such additional considerations as *suboptimal* factors that derogate from overall well-being. Accordingly, as Bernard Williams years ago observed, "defenders of such values are faced with the dilemma, of either refusing to quantify the value in question, in which case it disappears from the sum altogether, or else of trying to attach some quantity to it, in which case they misrepresent what they are about and also usually lose the argument."[3]

why can't we quantify?

Understood this way, the welfare-economic approach to environmental, health, and safety law seems capable of advancing normative views not through overt debate and suasion but through unremarked assumptions and exclusions. Most critically, as this chapter argues, the welfare-economic framework assumes that an uncontroversial basis exists for distinguishing between efficiency and equity, that economic valuation techniques can eliminate the need for direct political engagement with the protection of human health and the environment, and that public policy making can countenance only individually inscribed, preexisting interests, rather than interests that arise at the collective level or that only take shape in the wake of necessarily prior policy choices. These foundational assumptions work to obscure much of the reason behind publicly enacted environmental, health, and safety laws. Rather than representing an effort to actively enhance the position of workers, endangered species, future generations, and other beneficiaries of legal protection—to endow those entities with greater social value than they attract under current market and political equilibria—environmental, health, and safety laws instead appear to represent "irrational" overinvestments in protection. Likewise, rather than embodying a deliberate effort to use laws' inevitable culture-altering power to guide individuals over time to a mode of living that is safe, humane, and ecologically sustainable, environmental laws instead appear to embody "naïve" or "paternalistic" expressions of some unattainable or undesired future. In short, by resorting to a series of debatable techniques in order to maintain the perception that welfare maximization only passively and impartially aggregates the autonomous desires of individuals, welfare economists end up discarding many of the considerations that give meaning to environmental, health, and safety laws.

*

At its core, the welfare-economic policy-making paradigm offers individual welfare maximization as a positive account of human behavior and offers social or aggregate welfare maximization as a normative goal for the design of law and policy. This simultaneous endorsement of welfare maximization as both a predictive and prescriptive model does not render law irrelevant, because the paradigm carries with it important market-failure concepts according to which even perfectly rational individual actors will fail to maximize collective well-being. Thus, analysts tend to focus on situations of expected departure from textbook market conditions, such as incomplete or asymmetric information, negative externalities, public goods, collective action problems, or monopoly power. Policies are then designed to alter the decision-making environment of relevant individuals and entities so that behavioral incentives better align with the goal of welfare maximization. Often this task is simply thought to entail creating clear property rights or otherwise reducing barriers to economic exchange so that individuals may, in a decentralized fashion, improve overall well-being through their voluntary activities and exchanges. In extreme circumstances, market failures on the welfare-economic account may require more "interventionist" measures, such as imposing a tax on a particular activity in an amount equal to the level of harm that the activity is imposing on others. In that manner, the negative externality will be brought into the informational workings of the market so that only those activities generating benefits in excess of total costs will continue.

When it is not feasible simply to "get the prices right" in this manner, regulators themselves must attempt to determine optimal outcomes. The gold standard for such analysis is the Pareto optimality standard, which deems a policy proposal acceptable only if at least one individual is made better off by the proposal and no individual is made worse off. Further, according to the most stringent formulation of the Pareto criterion, all individuals *themselves* must decide whether they are made better or worse off by a proposal; in that manner, the welfare analyst can claim to have avoided altogether the need to compare well-being across individuals. Behind this aversion to interpersonal comparisons lies a familiar story about the intellectual origins of welfare economics: During the field's formative phase, welfare economists aspired to attain the level of scientific pedigree accorded to the natural sciences; thus, they sought to banish from the domain of welfare economics all modes of analysis that seemed ethical or otherwise nonobjective in focus, with comparisons of utility or well-being across individuals being the most obvious case. Armed with Pareto's seemingly unobjectionable standard of policy evaluation, welfare economists came to make strong claims about the positivistic nature of their discipline, as evidenced by the following characteristically assured example: "It should not be necessary to point out that, despite its slightly misleading name, the concept of a Pareto Optimum is completely objective and that our discussions are of a positive rather than a normative nature."[4]

Contrary to such claims, the Pareto standard is not objective, even in its most ideal formulation. Most obviously, the standard is built on the normative assumption that human individuals provide the only relevant metric for welfare analysis, as opposed to other interest holders, such as human generations, human communities, or even nonhuman life-forms. It also is premised on the notion that welfare analysis should be governed by measuring changes to the status quo distribution of rights and resources, as opposed to some socially imaginable alternative baseline. This latter assumption has far-reaching ethical consequences. As Amartya Sen has noted, because Pareto optimality does not permit inspection of the underlying distribution of initial entitlements, "a society or an economy can be Pareto-optimal and still be perfectly disgusting."[5] In addition to this normative complication, the Pareto standard also is not practicable, given the extraordinarily high bar it sets for permissible policy interventions. Very few proposals within complex and deeply interrelated societies, such as those of advanced industrial states, can avoid causing one or more individuals to be made worse off. Accordingly, the opportunity costs of the Pareto stance can be extreme: Even a policy that doubled the well-being of every member of society while making only one well-off individual slightly worse off would fail the stringent conditions of the Pareto test. Indeed, as Guido Calabresi has argued, once transaction costs and other real-world variables have been taken into account, there may in fact be *no* Pareto-improving policies available that have not already been adopted, at which point the Pareto standard becomes essentially "pointless."[6]

Largely for practical reasons, many economists shifted over time to a standard known as potential Pareto optimality or, alternatively, Kaldor-Hicks optimality. Unlike the Pareto standard, the Kaldor-Hicks welfare criterion requires only that the overall gains in well-being from a policy change outweigh losses, so that the "winners" could, in theory, compensate the "losers." Because this compensation is merely hypothetical in

Sen's critique of an initial position...

nature—that is, it need not actually be effectuated in order to deem a policy proposal efficient—the Kaldor-Hicks standard represents an abandonment of both the ideal of avoiding interpersonal welfare comparisons and of permitting individuals themselves to determine whether, and by how much, their well-being is altered by a policy proposal. Although efforts are still made to base welfare estimates on individuals' expressed or revealed preferences, the identification and interpretation of such preferences is far more complicated than in the Pareto context, because actual consent is no longer the standard of acceptability. In practice, the Kaldor-Hicks standard tends to prompt a shift from welfare maximization to wealth maximization, given that the dollar-weighted valuation of welfare impacts provides the most obvious and tractable method for operationalizing the hypothetical compensation test. In fact, some proponents of the economic analysis of law describe the Kaldor-Hicks efficiency criterion as being conceptually interchangeable with wealth maximization.[7]

Although there are important and sophisticated exceptions,[8] most uses of regulatory cost-benefit analysis fit this description, and, accordingly, most uses can be said to assume at heart that welfare maximization—as determined by the Kaldor-Hicks efficiency standard using conventional techniques of welfare estimation—has primary moral and political significance. Within the idealized cost-benefit state, no policy is passed that decreases aggregate welfare.[9] When regulators must set environmental, health, and safety standards, they use cost-benefit analysis to select the point of marginal equivalence between social costs and benefits, thereby maximizing the "bang for their buck" that citizens achieve from regulatory programs.[10] Similarly, by applying cost-benefit analysis in a serial fashion to an exhaustive range of existing and proposed risk-regulation programs, analysts can generate a list of policies ordered according to their cost-effectiveness, thereby providing society with a basis for making optimal use of the entire regulatory budget that it devotes to risk prevention.[11] Assuming agreement with the basic welfarist framework, neither the results of any single cost-benefit analysis nor the global priority-setting exercise should afford a basis for serious objection or dissent, as all important considerations will have been identified, tabulated, and weighted in a manner that leads to the identification of optimal outcomes.[12]

Recognizing how philosophically contingent this argument is, proponents of cost-benefit analysis in recent years have diversified their portfolio of justifications. They have argued, for instance, that cost-benefit analysis satisfies "both utilitarian and deontological accounts" of morality by forcing regulators to accept "people's actual judgments" concerning their own well-being, an argument that obviously assumes prevailing methods of identifying and measuring individual preference are an adequate proxy for "people's actual judgments."[13] Cost-benefit analysis also is said to serve important equity interests by helping society equalize the dollar amount spent per life saved across various agencies. Such an approach serves one conception of the principle of equal respect for persons by providing a uniform cost-effectiveness rate for society's efforts to prevent and manage diverse categories of risk. More generally, by providing comprehensive evaluation of the stakes involved in political choices, cost-benefit analysis is thought to offer important democracy-enhancing benefits, such as an improvement in the quality of political deliberation, a control for unreliable cognitive tendencies, a reduction in the distorting influence of interest groups, and an equalization of political power between traditionally dominant and traditionally underrepresented factions. Although often phrased in the conventional wealth-maximization terminology of cost-benefit analysis,[14] these equitable and democratic concerns need not be so narrowly conceived. Indeed, as other scholars have pointed out, advocates of cost-benefit analysis share much in common with the early twentieth-century Progressives who believed that the common good—and not just the aggregate satisfaction of preferences—could be furthered through improved conceptions and practices of rational government.[15]

These diversified justifications bring welfare-economic proponents out of a narrow utilitarian paradigm and require them, in turn, to entertain critiques from a similarly broad range of viewpoints. Most importantly, they must better defend the conceptual heart of welfare economics, which is its selection of the basic value criterion with which to measure policy consequences. This factor sets in motion the optimization calculus that determines, on the one hand, what will count as efficient and, on the other hand, what will be relegated to the realm of the overtly subjective, and therefore will be considered an "additional input" into policy making that, by definition, detracts from efficiency. As noted above, the dominant value criterion in welfare-economic-policy analysis is individual well-being, which typically is given concrete, tractable expression through the use of [willingness-to-pay estimations.] According to most welfarists, officials should pursue efficiency using these estimations and should disregard the distributive impact of policies. To the extent that fairness, sustainability, or other nonefficiency concerns militate in favor of particular outcomes, prominent welfarists argue that such goals should be pursued solely through the tax-and-transfer system, not through alteration of the substantive content of environmental, health, and safety laws. By focusing government policies exclusively on the maximization of efficiency in this manner, while letting other concerns be addressed through progressive taxation rates and transfer payments to the poor, officials can increase the "size of the pie" to its maximum potential, ultimately making it possible for everyone to be better off than they would be in a world in which nonefficiency concerns were permitted to directly influence policy choices.

nonefficiency concerns pursued through a tax-and-transfer system outside welfare economics...

*

Or so the story goes. Rather than an uncontroversial demarcation between technical and ethical domains of analysis, however, the conceptual division between efficiency and equity is itself an emphatically political gesture. It is the most basic reason why attempts to specify separate domains of positive and normative analysis in welfare economics fail and why environmental, health, and safety lawmaking can never be reduced to empirical technique in the manner hoped for by welfarists. When government officials state that "[t]he standard best practice in economic analysis is applying an approach that measures costs, benefits, and other impacts arising from a regulatory action against a baseline scenario of the world without the regulation,"[16] they are stating something more than a methodological conviction. Prioritizing the status-quo distribution of rights and resources is not a neutral or objective approach, any more than would be an approach using an entirely different distributive baseline. We could, for instance, demand that policy effects be measured using a distribution-sensitive welfare criterion, one that weighted more heavily gains to the worse off than gains to the well off. Alternatively, we could follow the lead of eminent welfarists, such as Amartya Sen, who offer an "objective list" of essential human goods or capabilities as the relevant indicia of social well-being that governments ought to pursue.[17]

More dramatically, as detailed in later chapters, we could value effects using prices that have been derived from a hypothetical market in which nonrenewable and depletable resources are placed under strict conservation measures for the benefit of future generations. In each case, we would derive efficiency calculations that are more consonant with the goals of equity and sustainability, so that the perception of an inevitable tension among our goals would be diminished.

The deep conceptual contingency of the welfare-economic framework can be concretely depicted in the context of water law, where efforts have been undertaken to replace existing allocation mechanisms with more market-based price and exchange mechanisms. Depending on levels of market competition and on the degree of identifiable differences among consumers, an unregulated water market might feature prices set in a discriminatory manner: So long as producers could practicably and legally price according to volume of usage, then the most vital uses of water—for drinking and sanitation—would be charged the highest per-unit rate, given the relatively high valuation that individuals would express for such uses. On the other hand, larger volume users—such as agricultural, industrial, and recreational users—would receive bulk discounts in recognition of the generally lower willingness-to-pay per unit of consumption that their activities would support. This market-allocation approach would likely have the salutary effect of reducing certain low-value uses, such as irrigation for ecologically inappropriate agriculture, which currently persist only through the support of public subsidization. However, by reducing public control over water pricing, the policy reforms of privatization and market allocation might also exacerbate problems of access for the poor.

Of course, proponents of market-based water allocation are not unaware of the problem of wealth inequality or of the need to guarantee access to basic drinking and sanitation water. Accordingly, they typically urge some form of publicly funded monetary assistance for the poor in conjunction with their proposals to introduce market-based allocation.[18] This approach, however, reflects the tendency of welfare economists to seek maximization of the overall value obtained from resource uses while allowing questions of distributive equity to be handled separately through some form of tax-and-transfer program.[19] The supposed trade-off between efficiency and equity is accommodated by allowing the market price of water to settle on its deregulated equilibrium and by transferring sufficient monetary resources to the poor to enable them to continue to participate in the market. Yet this effort to account for significant non-efficiency considerations through corrective devices such as transfer payments carries with it an unwarranted stigma: The language of "tradeoffs" inappropriately speaks of equity as a "cost" to be balanced against efficiency; it suggests a sacrifice of welfare rather than a promotion of life. This is particularly regrettable in the context of a fundamental resource such as water, where we would do better to recall the maxim expressed by Mary Wollstonecraft: "It is justice, not charity, that is wanting in the world."[20]

The perceived "tradeoff" between efficiency and equity can be replaced by an integrated standard of "equitable use"—a basic intellectual reform that eschews the separationist tendencies and value agnosticism of market liberalism in favor of an objective account of the basic or minimal determinants of well-being. For instance, in the field of water law, one demand-side management tool that has generated a great deal of attention and support among water experts recently is tiered or block water pricing.[21] Tiered pricing seeks to harmonize efficiency and equity in water usage, not by deregulating markets and transferring money to poor consumers, but by charging a low or even no-cost rate for the initial quantity of water consumed by a user and gradually increasing the rate as the level of consumption increases. This approach implicitly assumes that the most vital uses of water should be the most widely available and, indeed, should be accessible to all individuals without regard to their ability to pay. Generally speaking, larger-volume uses are less vital and therefore more in need of higher prices to incentivize conservation. In this manner, the efficiency calculus under tiered pricing becomes unhinged from willingness-to-pay as a welfare criterion; instead, volume of usage stands in as a proxy for socially determined use priorities. Accordingly, the efficient result—defined not in the sense of maximizing the satisfaction of monetized preferences, but in the sense of maximizing essential conditions for human survival and flourishing—also can be described as the equitable result.

Of course, this is not to suggest that efficiency is unimportant or that society should not seek to promote progress according to its chosen value metrics; it is only to suggest that the choice of value metric is an emphatically ethical and political exercise, one that should remain open to contestation and that should, at least occasionally, take a backseat to more direct discussion of matters of justice. The significance of keeping open this space for contestation and discussion can be appreciated through closer inspection of conventional welfare-economic valuation techniques. As noted above, the value criterion of individual preference, as approximated by willingness-to-pay, has come to dominate applied welfare-economic techniques such as cost-benefit analysis. However, because the cost-benefit procedure is utilized in contexts where individual market behavior by assumption has failed to ensure aggregate welfare maximization, it generally is not possible to *directly* observe individuals' preferences for the social tradeoffs involved in environmental, health, and safety decision making. Accordingly, economic scholars have devised various indirect measures of individual preferences for all manner of public goods, such as endangered species, worker safety, and visibility levels in national parks. Two broad categories of such methods are the revealed-preference methodology, which seeks to glean implicit valuations from actual market choices that are related in some fashion to public goods, and the expressed-preference or contingent-valuation methodology, which attempts to provoke individual market evaluations of public goods through experimental surveys. Of the two categories, revealed-preference methodologies are thought to produce especially realistic valuations because, unlike hypothetical survey instruments (or political referenda, for that matter), the choices examined occur under the disciplinary constraints of actual markets with real economic resources at stake.[22]

Behind such methodologies lies a conviction that individual behavior can be best described and predicted by assuming that individuals seek to maximize their interests through the use of instrumental means-ends reasoning. Individuals are believed to have stable preference functions and to act rationally in the pursuit of maximal satisfaction of those functions, so that their behavior can be taken as a reliable empirical source of information about their interests. For instance, researchers have sought to find an indirect measure of a parent's monetary valuation of her infant's life through her choice to pay higher prices for organic baby food, which presumably contains fewer pesticide residues and other potential threats to the infant's well-being.[23] As this example makes clear,

the analyst must always rely on a series of presuppositions to link observed behavior with the ultimate conclusion that individuals have revealed their valuation for the "good" in question. Despite the oft-heard
claim that such studies represent "*in fact* reflections of individual preferences, and hence utility,"[24] the interpretation of observed behavior is actually a slippery exercise in which the analyst must adopt contestable
assumptions concerning the opportunities, information, and choice criteria that confront observed individuals. At times, these assumptions
appear to rest more on the personal beliefs and professional customs of
the analyst than on sustained engagement with the actual circumstances
of observed individuals. As Mark Sagoff puts it, "Choice is at best a conceptual construct inferred from ad hoc descriptions of behavior—
descriptions that themselves presuppose beliefs about available options
and therefore about preferences."[25]

For instance, the values being assessed through welfare-economic-
policy analysis often take the form of opportunities that would be lost if
a policy is adopted. Typically, this attempt to conjure and quantify what
"might have been" is as much art as it is science. This is particularly the
case when considering vastly complex and dynamic systems such as the
myriad social and natural systems driving, and being driven by, climate
change. In one prominent cost-benefit analysis of climate change, for
instance, the author's decision to add a variable representing enhanced
recreation opportunities in a warmer world resulted in monetized benefits that tended to swamp the impact of estimated morbidity and mortality.[26] In essence, the fact that the analyst could identify easily monetizable
benefits that would accrue to affluent individuals from, for instance, being able to mountain bike an extra week each year became an inordi-
nately significant driver of the cost-benefit assessment. Apart from being
unduly wooden—why assume, after all, that individuals will continue to
value mountain biking at the same amount if they know that their extra
days of recreation have been funded through an increase in malaria
among the equatorial poor?—the exercise also was open to considerable
selection bias. In a real sense, climate change has the potential to affect
every natural and social system on the planet, systems about which our
present understanding is highly incomplete and imperfect. To choose
among such effects and offer the resulting calculation as objective is
convincing only because, one suspects, the ritual comports with a deeply

ingrained desire to imagine that our most difficult policy choices can be reduced to scientific or technical terms.

Conceptual problems of this sort remain unaddressed even within the most well-developed revealed-preference literature to date. When evaluating environmental, health, and safety regulations, U.S. federal agencies typically use a value of human life that has been empirically derived through observations of individual behavior in labor markets. The central premise of these studies is that, other things being equal, workers demand a higher wage in order to accept a higher level of occupational fatality risk. Large data sets have been analyzed with this assumption in mind, and a consensus view among practitioners has developed that the implicit value of life revealed by wage premiums lies somewhere around $7 million.[27] Such figures now are routinely used to compute the monetized benefits of proposed environmental, health, and safety regulations, affording in the process a ready basis for comparison against the estimated economic costs of regulation.

The veracity of the wage premium literature depends critically on the tightness of fit between subject labor markets and the competitive market ideal. At least two of the most important assumptions of the market ideal, adequate risk awareness and free mobility among laborers, seem questionable in this regard. Even in the theory, these assumptions are resistant to clear specification. After all, what precise level of information and understanding, what exact degree of opportunity and mobility must be afforded to individuals before we deem their acceptance of a mortality risk "voluntary"? Moreover, the assumptions present a muddy empirical picture once we closely examine market operations.[28] Ironically, many data problems arise in the wage-risk-premium literature precisely because the United States managed to drastically reduce its occupational hazard levels over the course of the past century, an accomplishment that resulted in substantial part from protective legislation that did not depend on cost-benefit analysis for its justification.[29] For this reason, the few remaining segments of the economy that exhibit an occupational mortality rate high enough to support the wage-risk-premium methodology also tend to be segments populated by those with the least social, economic, and political capital—variables that themselves may be expected to influence the opportunities, risk awareness, and preference orderings of observed individuals.[30] Thus, what the cost-benefit practitioner regards as choice (and hence preference—and hence utility) may better reflect the analyst's inclination to treat preexisting power relations and other features of employment markets as normatively privileged. According to some studies, for instance, the implicit value of life revealed by wage-risk interactions appears to be several million dollars higher for union workers than for nonunion workers.[31] Union membership, in other words, seems to be a strong determinant of the size of the compensating wage differential—a wage-risk effect that more closely captures bargaining power than willingness-to-pay. Assuming the accuracy of this data, the analyst faces a difficult question: Which set of workers—those endowed with collective negotiating powers or those not endowed—should supply the relevant societal "preference" for investing in safety? The answer is not to be found in cost-benefit analysis but instead in sustained ethical inquiry, lest we enshrine into public policy aspects of our economy that we regard as regrettable side effects of market capitalism, rather than as desiderata in their own right.

Nonetheless, proponents of cost-benefit analysis generally believe that wage-risk-premium studies do provide a sufficient basis for assuming consent to the imposition of all manner of health and safety risks, a belief that in turn leads them to argue that cost-benefit aggregation does not involve human life or health at all, but instead only the "monetary equivalents" of such values.[32] Two important analytical leaps are implicit in this claim: first, the decision to treat apparent valuations of risk as scalable so that, for instance, a willingness to invest one thousand dollars to avoid a $1/1,000$ risk of death can be converted into a uniform value of one million dollars per statistical life; and second, the decision to utilize valuations derived from the blue-collar labor market in all manner of cases in which the government acts to reduce threats to life. To the extent that proponents of interest aggregation acknowledge these analytical leaps, they often defend the leaps by noting that most human health risks from environmental hazards are quite small and that, therefore, officials can safely assume individuals would consent to the risks for a price comparable to the compensating wage differential.[33] This defense is inadequate for two reasons. First, as the psychological literature on risk perception demonstrates, individual responses to even actuarially identical risks vary dramatically based on the risks' qualitative characteristics, and thus officials cannot be said to respect "people's actual judgments"[34] when they extrapolate from the wage-risk-premium literature to other contexts. Second, a variety of adverse health risks associated with pollution and other hazards are not trivially small, and thus it is unwise to assume that the monetized value of such hazards can simply be determined as a linear function of the compensating wage differential.

Cost-benefit analysis's limitations in this respect seem to be driven by the procedure's purely individualistic conception of value. Without an identity—and therefore without a willingness or even an ability to pay for protection—those lives that are threatened by statistical risks seem not to represent human lives at all, but instead only their "monetary equivalents." Statistical risks, however, represent "none of us" and "all of us" at once. Thus, the full import of statistically identified harms only appears visible—only emerges—when evaluated from the perspective of the same entity through which the data were derived, that is, from the perspective of "all of us." Because interest aggregation approaches like cost-benefit analysis refuse to see "all of us" as interest holders, they struggle to treat environmental, health, and safety regulation with the moral richness that the field deserves. Indeed, as Lisa Heinzerling notes, by pricing human life and sanctioning actions that place it in jeopardy in advance of those actions occurring, "the most basic kind of right—the right to be protected from physical harm caused by other people, on equal terms with other people—is denied to those whose lives are framed in statistical terms."[35] As a consequence, we tolerate treatment of statistical victims that would be considered well nigh criminal in ordinary contexts. We do this simply because the body and the fingerprints necessary to establish the crime become invisible once they are reduced to individualistic shares. From the perspective of "all of us," air pollutants cause thousands of premature deaths each year in the United States; from the perspective of any particular victim, however, no cause and no perpetrator can be clearly identified and, thus, "none of us" are ever seen to be harmed.

Better alignment is needed between the nature of the harms we suffer and the techniques of valuation we deploy. Welfare economists assume that the proper measure of the worth of a public good is the amount that individuals would be willing to pay in order to preserve the good in a well-functioning market. A different approach altogether would be to assume that such valuations are more reliably captured through society's willingness to act collectively in order to preserve the threatened good. One could, for instance, simply allow individuals to express their preferences through democratic channels, by voting in favor of environmentally inclined or disinclined politicians or by holding direct referenda on environmental policies.[36] As Sagoff observes, unlike the comprehensive rationality of cost-benefit analysis, "[t]his kind of rationality depends on the virtues of collective problem solving; it considers the reasonableness of ends in relation to the values they embody and the sacrifices we must make to achieve them."[37] To be sure, a monetary value of human life will be implied after the fact by the amount of resources necessary to fulfill the aims of protective policies. But this value will not have driven the initial selection of policy; instead, the value will simply be an ancillary effect of a policy choice that was premised on social values, explicitly discussed and mediated through democratic decision-making processes.

Rather than embracing the view that collective harms require collective valuation, however, proponents of cost-benefit analysis are heading in precisely the opposite direction, promoting the goal of a perfectly "disaggregated" value of life.[38] This process of disaggregation would begin by allowing regulators to vary the value of a statistical life based on demographic characteristics such as race, gender, or age. Ultimately, regulators would aspire to set policies according to unique valuations offered by whichever *specific* individuals would be affected by a given decision, perhaps through online studies conducted in connection with each new rule making. In practice, these steps toward disaggregation would mean that certain racial minorities and other economically disadvantaged groups would tend to have a lower value of life assigned to them than under the current practice of using a uniform value of life. Such seemingly discriminatory government treatment does not concern Sunstein, one of the chief proponents of disaggregation, because in his view the government would merely be acting as a passive instrument of optimization, rather than as an agentic entity that might be charged with responsibility for the categories that it constructs. In Sunstein's view, if race-based differences in the value of life were used for policy judgments, they "would not be the result of a government decision to take racial characteristics into account; in fact [they] would not be a product of any kind of group-level discrimination on the government's part."[39] Instead, disaggregated life valuations would simply ensure that the government acts like any other seller in the marketplace, giving people only the level of protection that they actually want in light of their existing preferences and budget constraints.[40] This conception again arbitrarily privileges the status-quo distribution of rights and resources in the economy, treating the content of environmental, health, and safety law as determined by the status quo, rather than as a part of the backdrop of laws and regulations that determine it. This is a somewhat astonishing concession: Even in the context of discrimination by private actors, our laws typically do not allow defendants to use the discriminatory preferences of their customers as an alibi for unethical conduct.[41] Yet this alibi for discrimination is precisely what Sunstein wants to grant to the government.

The distinction between [outcome-determining] and [outcome-determined] views of the content of law is perhaps best illustrated through analogy to a context in which the sublimation of law to welfare analysis is more obviously disconcerting, or at least unfamiliar. In the wake of the events of September 11, 2001, the Office of Management and Budget (OMB) issued a call for research on how to measure the value of liberty and privacy interests that would potentially be sacrificed by new antiterrorism measures.[42] Responding to this call, researchers at Harvard attempted to measure individuals' willingness to trade off civil liberties for safety and convenience during airport-security checkpoint procedures. In particular, researchers questioned survey respondents' willingness to accept racial profiling in exchange for reduced time spent waiting in line. Not surprisingly, white respondents were generally more willing to accept racial profiling than nonwhite respondents.[43] The principle of race neutrality did figure somewhat in the white subjects' decision making: When told that only nonwhite groups would be targeted for profiling, white respondents expressed an unwillingness to support the explicitly race-based program for a meager ten-minute time savings. For a thirty-minute savings, however, the white respondents overcame their moral qualms.[44] They revealed, apparently, their willingness to pay to preserve the principle of race neutrality.

By calling for empirical research of this nature, the OMB sought to develop procedures for estimating the welfare impacts of civil-liberties

[margin annotation: outcome-determining vs. outcome determined ...]

restrictions. From the welfarist perspective, the value of civil liberties would thus be made contingent on the level of burden that individuals are willing to tolerate in order to maintain them. In a constitutional scheme of government, however, the cart lies at the other end of the horse: Civil liberties are protected by law, and their value, assuming one insists on quantification, is implied by the level of burden that individuals are made to accept in order to maintain them. At least on their face, environmental, health, and safety laws also frequently display this structure: Employees are entitled to a safe workplace, endangered species to necessary habitat, citizens to clean air and water, and so on. These entitlements are meant to be inviolable; that is, they are meant to be protected by law, so that their value is a creation of law, rather than a determinant. Just as a constitutional scheme of rights is intended to protect certain interests of individuals from incursion by majoritarian politics, democratically enacted environmental laws seek to protect certain interests from shortsighted destruction by land development, pollution, and other potentially harmful activities. In both cases, we seek to enshrine the collective values and aims of our better selves, knowing that our lesser selves periodically will be tempted to waver, shirk, or cheat—that we might, for instance, find ourselves willing to accept the racial stigmatization of others just to save time in a security line.

value of right is a creation of law, not a determinant

Not sure how they are still valued then...

*

The possibility of collective valuation through traditional political channels suggests a final limitation with the interest-aggregation model of cost-benefit analysis. Contrary to the standard assumption of many schools of policy analysis, markets and other sociolegal systems are not characterized by linear, unidirectional relationships between public values and the law, on the one hand, and between the law and public behavior, on the other. Rather, the system of relationships among these forces contains feedback loops, oscillations, and other characteristic traits of complex adaptive systems. Thus, just as legal policies may affect behavior in complex and unanticipated ways, so too may policies alter the beliefs and attitudes of individuals, including perhaps the very beliefs and attitudes that initially justify a policy choice according to welfare-economic analysis. As Sunstein pointed out in an early influential article, much of environmental law and policy making can be expected to have these kinds of endogenous effects.[45] More broadly, the foundational nature within any market economy of energy, natural resource, agriculture, pollution, transportation, land use, housing, and other infrastructure policies means that governments simply cannot abstain from influencing—directly or indirectly—the content of individual preferences. Because cost-benefit analysis takes those preferences as its starting point for policy making, the methodology captures only a static picture of what is unavoidably a dynamic process.

static picture of a ✓ *dynamic process*

As noted above, the cost-benefit procedure typically expresses a political conservatism, in that it takes as given the status quo distribution of rights and resources. It also generally displays a conservative cultural bias, in that the normativity of collective action is made to depend on *previously* identified preferences, rather than on preferences that result from the exercise of collective engagement. Under welfare-economic analysis, no allowance is made for the possibility of citizens to adapt their preferences in light of changed circumstances, to acknowledge the ethical responsibility created by past actions, to accept as being well and just any newly imposed constraints on their harmful activities, or simply to get on with the project of life by deriving welfare in new—but not necessarily inferior—ways. This lens is unduly narrow: Just as certain attributes and behaviors of complex, adaptive biophysical systems cannot be predicted by examining individual system components alone, certain values and aims of the "social organism"[46] cannot be identified or predicted through the simple aggregation of atomized preferences or interests. Instead, those values and aims in part emerge through the operation of social institutions and procedures themselves.[47]

As noted above, policy making that is overtly collective in orientation is thought to carry a risk of "insulting [people's] dignity" by rejecting their "actual judgments."[48] Yet there is insult also in attributing meaning and significance to behavior that individuals themselves may not desire or intend. Because the economists' methods of identifying preferences necessarily entail subjective moments of interpretation and judgment by the economist himself, this danger is quite real. Moreover, for many pressing policy matters, individual preferences are likely to be nonexistent or ill formed in the absence of an appropriate forum for discussing and determining social goals. To give just one example, we will not know our preferences with respect to using cloned or transgenic livestock for human consumption unless and until we have a body of

ill-formed preferences ✓

relevant experience from which to draw upon in our evaluations. Such experience will only occur with the prior consent of our political community, whether actively or passively granted. Thus, it makes little sense at present to hinge policies regarding cloned or transgenic livestock on a contrived depiction of our existing wants. What seems required instead is some mechanism for discussing what it is that *we want to want*.[49] Questions of that nature depend not on cost-benefit balancing, but on the will of a community to conceive of and to realize a transformation of culture toward some shared ideal.

"what we want to want"

Defenders of welfare economics would be quick to argue that such deliberation can and should take place on an individual level and in other private forums, but not within government policy making. In the liberal political tradition, they would argue, the government's responsibility is limited to providing optimal enabling conditions for individuals to pursue their life visions by themselves, for contemplating what they want to want on an individual basis with minimal intrusion from the state. The preference-aggregating approach of welfare economics takes this liberal aspiration to its extreme, by representing state action itself as [ministerial] and [nonagentic] As chapter 2 demonstrated, however, such attempts to efface the collective are conceptually problematic in that they are incapable of accounting for their own normativity; that is, they fail to identify the agent that decided, through adoption of the welfare-maximization mandate, to deny its own agency. Even within the thinned-out framework of welfare economics, the normative dimensions of social choice continually resurface: What baseline should be chosen as the standard against which welfare consequences are "impartially" assessed? What method of quantitative valuation, if any, should be selected for policy effects that implicate life and other sacred "goods"? Should worker safety, endangered species, and civil liberties be treated as contingent resources or as constitutional givens? What degree of self-consciousness should agencies and government officials hold regarding the fact that policy making never merely traces individual preferences, but always also is at work influencing their content? Even if normative agreement on such questions could somehow be definitively achieved—even if, for instance, the apparent intractability of such questions could be tamed through universally embraced modeling assumptions—the interest aggregation framework still would imply the existence of some political collective, the agency of which was exercised in choosing the assumptions.

The most worrying danger presented by cost-benefit analysis is not that we will choose the wrong modeling assumptions, but that the full power and responsibility of our collective agency will become lost amidst the rhetorical force of an interest-aggregation exercise that purports to take account of all relevant consequences of social choice. This semblance of comprehensiveness is misleading: The answers that cost-benefit analysis provides work a narrowing of the questions that environmental law asks.

PART THREE THE ENVIRONMENT OF THE OTHER

CHAPTER FIVE

Other States

LAW CONTAINS ITS OWN GEOGRAPHY. Implicit within the marbled layers of local, state, federal, and international rules that comprise any country's environmental law regime is a vision of what the world looks like, how its territories are differentiated, how they relate to one another, and whether they are surpassed by forces greater than their sum. The geography implicit in law is often strange, even to lawyers. Many U.S. environmental laws, for instance, do not suggest on their face that there is an environment beyond the nation's territorial borders. Instead, the geography of U.S. law reflects the traditional Westphalian conception of sovereignty, in which each individual state is deemed to have nearly absolute authority over the space within its physical borders. States thus depict themselves, in their laws, as both politically autochthonous and, somehow, ecologically autonomous. Apart from certain recognized sites of common heritage, such as Antarctica, outer space, and the deep seabed, and apart from certain pervasive media, such as the vast international waters within which political territories are to be found, the starting principle of environmental law is that "States have . . . the sovereign right to exploit their own resources pursuant to their own environmental and developmental policies."[1]

Rio Declaration

Occasionally, these hermetically sealed nodes of legal authority are recognized to be interconnected through paths of environmental impact that give rise to limited bilateral or regional agreements, such as the series of treaties that have structured relations between the United States and Canada with respect to North American air pollution and regional management of the Great Lakes. Although limited in practical effect, these agreements do represent an effort to implement the often-forgotten corollary to environmental law's baseline condition of Westphalian sovereignty, namely, that "States have . . . the responsibility to ensure that activities within their jurisdiction or control do not cause damage to the environment of other States or of areas beyond the limits of national jurisdiction."[2] In some instances, the environmental laws of the United States and other countries have gone even further to recognize problems of a truly global scale, problems that demand an integrated, multilateral response. Among such cases, the legal regime to arrest the production and consumption of ozone-depleting substances is often heralded as a particularly effective example of international environmental lawmaking, having achieved nearly universal endorsement and having contributed to a dramatic decline in the use of such substances during its first two decades of existence. Accordingly, much of the agenda of promoters of international environmental law at present is to expand the list of problems that are recognized, like ozone depletion, to be global in nature. The hope of these advocates is that the geography implicit in law will over time come to resemble that of the earth sciences. As the category of environmental pathways acknowledged by law expands and diversifies, and as the operations of those pathways come to be seen as hemispheric or global in scale rather than national or regional, then eventually the claims of deep interconnection so prominent in environmental science, and so urgently pressed in environmental politics, also will find concrete expression in environmental law.

In recent years, the United States has come to be seen as a serious impediment to this integrative agenda, evidenced most prominently by the nation's withdrawal from the Kyoto Protocol but also apparent in the U.S. stance on genetically modified agriculture, persistent organic pollutants, and other international environmental issues. This widespread perception of U.S. recalcitrance is striking when juxtaposed against the commitment to international cooperation that once was demonstrated by the nation's environmental statutes. The United Nations Environment Program Participation Act of 1973, for instance, declared that "[i]t

Geography of the law will continue to expand...

is the policy of the United States to participate in coordinated international efforts to solve environmental problems of global and international concern."[3] Earlier, in 1970, the U.S. Congress "commend[ed] and endorse[d]" an effort of the International Council of Scientific Unions and the International Union of Biological Sciences to study "one of the most crucial situations to face this or any other civilization—the immediate or near potential of mankind to damage, possibly beyond repair, the earth's ecological system on which all life depends."[4]

Both of these statutes pledged not only moral but financial support to the international community, as did amendments to the Foreign Assistance Act adopted in 1977. These amendments began with a congressional finding "that the world faces enormous, urgent, and complex problems, with respect to natural resources, which require new forms of cooperation between the United States and developing countries to prevent such problems from becoming unmanageable."[5] In light of these problems, the amendments directed the president "to provide leadership both in thoroughly reassessing policies relating to natural resources and the environment, and in cooperating extensively with developing countries in order to achieve environmentally sound development."[6] Other examples of U.S. efforts to assert international environmental leadership include the Federal Water Pollution Control Act, which instructed the president to take those actions necessary to ensure that other countries reduce water pollution even within their own borders, and the Ocean Dumping Act, which directed the secretary of state to promote effective international cooperation to protect the marine environment.[7] This mode of national self-awareness regarding international environmental responsibility was also encouraged under some U.S. federal court interpretations of NEPA[8] and under a 1979 Carter administration executive order,[9] all of which encouraged consideration of the extraterritorial environmental impact of major actions by U.S. governmental actors.

Like U.S. environmental law more generally, these various efforts to promote internationally cooperative arrangements received strong nonpartisan support at the time of their adoption but have since tended to languish amidst the politicized and polarized atmosphere surrounding U.S. environmentalism since the beginning of the 1980s. Although numerous factors have contributed to this decline, this chapter argues that global environmental law has suffered particularly from the expanding influence of the economic reform movement, given that the instrumentalism at the heart of that movement is incapable of adequately accommodating intersubjective relations between states. In contrast to the self-consciousness demonstrated in early federal environmental statutes—in which the United States is depicted as a subject with responsibilities to foster and lead international dialogue among other political subjects concerning environmental protection—the risk-assessment-and-cost-benefit-analysis framework aims specifically to deny the U.S. political community a view from within itself. As argued in chapter 2, advocates of economic reform ask policy makers and bureaucrats to, in essence, regulate from nowhere, as if they perceive and respond to environmental policy issues from an impartial viewpoint in which the fact of the government's identity, relationality, and responsibility is ignored.

As this chapter argues, the welfare-economic paradigm does not provide an adequate vehicle for harmonizing the ideal of an impartial and comprehensive welfare analysis with the reality of state-centric policy making, which depends always on a view from within. In the absence of a truly global cost-benefit state, the economic framework still must consider transboundary impacts—whether caused by or caused in a state—yet the framework lacks resources for determining how to consider them. With respect to the view outward, the economic framework denies the political community an adequate vantage point for recognizing the ethical significance of its actions and for appreciating the need constantly to consider its responsibilities to others, even when fashioning environmental laws that might traditionally have been considered to fall within the domain of the state's sovereign prerogative.[10] How, after all, *are* the interests of individuals abroad to be treated when tabulating welfare consequences of domestic policy? For some, the utilitarian aspiration to impartiality and comprehensiveness should lead to a global welfare calculus in which the political affiliations of individuals are ignored altogether[11]; in practice, however, economic analysis must always consider extraterritorial effects with reference to the specific political community that has turned to cost-benefit analysis for aid. Thus, the question of how to regard foreign individuals cannot be resolved through the optimality logic of welfare economics alone, since the question goes to the constitution of the optimization model itself, to the foundational decisions of what interests and on what basis to optimize. With respect to the view inward, practitioners of cost-benefit analysis also must decide whether to treat the actions of other states that cause harm domestically as dependent or independent variables. In other words, should other political communities be treated as nonagentic systems, modeled and predicted like hurricane cycles, dose-response curves, and fishery yields, or should they be greeted as subjects of reason and mutuality? Adopting the former approach would be consistent with welfare economists' aim to formalize policy making as a deterministic, objective exercise. At the same time, however, it would discourage the state from seeking cooperative relations with other sovereigns whose activities increasingly affect the ability of regulators to achieve even domestic environmental goals, quite apart from the question of whether the interests of non-nationals should be included within those goals.[12]

To sharpen understanding of these dilemmas, this chapter examines two flawed strategies by which the economic approach to environmental law has attempted to accommodate the field's inevitable transnationality. First, regulatory isolationism in the United States has been defended on the grounds that environmental sustainability is best attained only after nations achieve sufficient economic growth to afford the "luxury" of protecting the environment and that, in turn, the best way to promote economic growth is through trade liberalization and other neoclassical reforms that feature less rather than more regulation. Second, to the extent that internationally cooperative regulation has been acknowledged as necessary, it has been modeled as a strategic game in which each state cares only about maximizing its own citizens' aggregate well-being. From this perspective, international law and international relations are seen merely as spaces within which rationally motivated competition occurs, rather than as forums for discussion and mutual social influence. This chapter argues that these approaches should be rejected: Rather than ignore the environmental pathways that bind states together in deep interdependence regardless of domestic economic conditions, rather than simply measure and accept the behavior of other political actors as an empirical given when establishing environmental policies, rather than reduce the self-understanding of international law to a silent—and all too frequently destructive—game theoretic problem, the state instead should engage its sovereign counterparts in reasoning toward shared environmental goals, a dialogic potential that once was clearly recognized and embraced by U.S. environmental law but that now seems obscured by the pervasiveness of [economic instrumentalism.]

*

Before addressing the welfarist approach to international law, it is necessary first to address the argument of those who contend that international regulatory efforts in protection of the environment are simply unnecessary. To some proponents of the economic reform agenda, the need for global environmental regulatory cooperation is vastly overstated, given their conviction that the ends sought by environmentalists can be best attained through relatively undirected, wealth-driven societal changes. Their conviction has some empirical support: Beginning in the early 1990s, economic researchers demonstrated a relationship between economic growth and environmental quality known as the Environmental Kuznets Curve (EKC), a finding that quickly came to stand for the proposition that nations tend to "grow their way to sustainability." The EKC refers to an inverted-U relationship in which environmental degradation tends to increase in the early stages of industrialization but then reverse course as countries achieve higher levels of per capita income. This fairly intuitive story appears first to have been empirically demonstrated in a World Bank study finding that particulate matter and sulfur dioxide pollution in forty-eight cities around the world peaked at income levels around five thousand to eight thousand U.S. dollars per capita and subsequently declined at higher income levels.[13] A large number of additional studies have replicated this result using other pollutants, time frames, and geographic locations.[14] In turn, numerous commentators have relied on the data to argue that environmental problems in the developing world can only be ameliorated in conjunction with economic growth. Occasionally, they have made the more-zealous claim that such growth by itself will provide the necessary catalyst to invest in environmental quality.[15]

These commentators are right to find a link between poverty and environmental degradation and, therefore, to cite protection of the environment as an additional reason to encourage appropriate economic development in poverty-stricken nations. However, as subsequent research has made clear, they ignore a variety of factors that render reliance on the EKC problematic as a simple remedy for pollution, aquifer depletion, deforestation, and other environmental problems. First, and most notably, the EKC simply has not been demonstrated for many actual or potential sources of environmental harm, including greenhouse gas emissions, municipal solid-waste generation, and use of potentially toxic pollutants.[16] Moreover, some sources of pollution that eventually do decline per unit of output nevertheless continue to increase in terms of aggregate emissions.[17] To the natural resource or ecosystem service being affected by a pollutant, only the latter variable matters. It also bears noting that some forms of environmental harm, like biodiversity loss, are irreversible. Thus, the redemptive half of the inverse-U curve does not occur for those species of flora and fauna that are sacrificed during a state's transitional development phases. Finally, a great many environmental issues involve cases of transboundary harm. These cases confound the atomistic conception implicit within the EKC argument, according to which each country's decisions regarding the perceived tradeoff between industrial activity and environmental quality are thought to affect only its own citizens.

Species decline is irreversible ...

For all of these reasons, the developed world has an interest in helping both to flatten and to shorten the EKC of developing countries, promoting active regulatory efforts that go beyond mere pursuit of economic expansion. An even deeper potential problem with the EKC-based argument is that, by assuming what has been true for the developed world will invariably become true for the developing world, commentators may have committed the fallacy of composition. That is, the environmental improvements witnessed in countries like the United States and captured by the EKC data may have occurred not simply because the environment is a luxury good that only a wealthy society can afford, but also because the United States has been able to "export" some of its environmentally degrading activity to other nations. If externalization of this nature has occurred broadly across environmental measures, then the EKC experience will not be as readily generalizable as commentators want to believe: Once we refuse to treat environmental costs as external, globally dispersed, or otherwise off the balance sheet, elementary principles of logic will prevent all nations from becoming net exporters of environmental degradation. An important study of forest preservation supports this hypothesis. Using a system-dynamics analysis that models, inter alia, deforestation rates, domestic GNP, and the GNP of a country's trading partners, Corey Lofdahl has shown that between 1976 and 1991, "an increase in either the percentage trade with a high GNP trading partner or an increase in a traditional trading partner's GNP results in decreased local forest area for the country in question."[18] In other words, economic growth in developed countries appears to be correlated with deforestation in developing countries. Thus, when commentators argue that "the primary solution to [tropical deforestation] will be higher growth and a better economic foundation so as to secure the countries concerned the resources to think long-term,"[19] they overlook the possibility that favorable reforestation trends in the developed world may actually depend upon contrary trends in less-developed tropical states. The EKC experience that is depicted as universally replicable may in fact be contingent on the availability of other nations whose resources are open to international exploitation.

*

Opening the world's resources to market exchange is precisely the aim of trade liberalization, a goal espoused with relatively little qualification even in "green" documents such as the Rio Declaration on Environment and Development.[20] At the heart of the case for trade liberalization is an efficiency argument in the spirit of Pareto optimality, one that purports to show that states can improve overall well-being by lowering trade barriers, without making any state worse off. David Ricardo's famous law of comparative advantage posited a case in which one country (Portugal) had an absolute efficiency advantage over another country in the production of two different goods (wine and cloth), but the second country (England) had a comparative advantage in one of the goods (cloth).[21] Thus, because Portugal's wine was worth more to England in cloth than the amount at which Portugal could produce cloth, both countries stood to gain from specializing in their respective goods and permitting free trading across their borders. In that manner, the two goods could be produced in amounts sufficient for both countries' consumption, but with far greater efficiency than when produced separately in both England and Portugal. This seemingly unimpeachable economic logic—which continues to form a key element of the argument in favor of liberalized trade—also now enjoys strong empirical support, as vast increases in global production over the past fifty years are attributable, in substantial part, to the liberalization of international trade that has been fostered by the General Agreement on Tariffs and Trade (GATT); its successor institution, the World Trade Organization (WTO); and numerous other trade agreements.

In developing his argument, however, Ricardo was careful to point out that the law of comparative advantage does not hold within a single state and, indeed, only works between states due to the international immobility of capital. Specifically, he noted that "[t]he difference ... between a single country and many, is easily accounted for, by considering the difficulty with which capital moves from one country to another, to seek a more-profitable employment, and the activity with which it invariably passes from one province to another in the same country."[22] If instead capital could flow freely across borders, then the global situation would become no different than the provincial: Capital would flee England for Portugal, where it would take advantage of absolute, rather than merely comparative, advantages in the production of wine and cloth. England would gain the benefit of cheap prices for wine and cloth imports but simultaneously would lose the employment base that helps to support a market for them. Despite Ricardo's admonition, this assumption of capital immobility has been largely ignored by contemporary theorists and policy makers, even as levels of international finance and investment have expanded exponentially. This is a curious omission: Just as awareness of the fundamental ecological interdependence of the planet has challenged the Westphalian concept of territorial sovereignty, the increasing permeability of national borders to capital migration should upset the notion of universally desirable free trade. As famed economist Paul Samuelson put it, the assumption of "zero net capital movements" underlying the conventional free trade model "smacks of Hamlet without the Gloomy Dane."[23]

Thus, the critical footnote to the comparative-advantage theory in a world of internationally mobile capital is that states are not necessarily all better off—or even no worse off—under liberalized trade, as is commonly argued. Instead, only the aggregate wealth of countries is certain to be enhanced, while any particular country may come out ahead or behind. To be sure, a host of theoretical and empirical complications are being ignored in this quick description. For instance, even as England loses employment associated with wine and cloth production, it gains in the form of increased returns to the owners of capital that might, in turn, be redistributed to laid-off workers via the tax-and-transfer system.

[margin handwritten note: Because capital can flee to take advantage of absolute, rather comparative advantages.]

Whether and to what extent international capital mobility has also undermined this "nationality" aspect of capital—that is, its amenability to enlistment in service of domestic tax-and-transfer policies—is an issue worthy of serious attention. Similarly, the simplified two-goods economy of Ricardo's example may understate the extent to which England can diversify its production to remain an attractive location for investment even in a context of internationally mobile capital. Whether absolute production advantages are distributed sufficiently evenly among nations to render the capital-mobility objection moot in this manner also merits further inquiry. Finally, the fact that overall welfare gains may be expected to flow from appropriately regulated trade liberalization remains an important argument in its favor, albeit one that requires much more vigilant qualification than it tends to receive in free trade debates. Those who believe that the distributive impact of economic activity deserves close attention within policy formation cannot ignore the capital mobility wrinkle in the way that the most ardent proponents of free trade have. Nor can they be satisfied in this context with the conventional welfare economic response of relegating equity considerations to the tax-and-transfer system. In light of the absence of strong representative institutions of redistribution at the global level—and in light of the relatively meager amount of voluntary redistribution that tends to occur in the place of such institutions—they must feel at least somewhat ambivalent toward trade liberalization when it occurs under conditions of international capital mobility, an ambivalence that would place them in the illustrious company of not only Ricardo and Samuelson, but also Adam Smith and John Maynard Keynes.[24]

In addition to this distributive qualification, some argue that free trade in an economy of internationally mobile capital may be in tension with the goal of environmental sustainability. Most prominent among these concerns is the familiar fear that international competition for capital may cause an unraveling of national production standards, including labor, environmental, and social safety-net standards. Because production cost is not simply a function of labor hours, as was believed in Ricardo's time, but rather a function, inter alia, of wages, taxes, and regulatory burdens, internationally mobile capital can be expected to flock to states that offer the most attractive package of inexpensive but productive labor, low taxes, and lax regulations. This "race to the bottom"

argument is hotly contested, both as a theoretical and an empirical matter. Some have argued, for instance, that equalization of regulatory standards eventually will occur upward as more states undergo the process
of economic development and experience its tendency to increase public
demand for environmental amenities and to decrease dependence on
environmentally intensive industries.[25] To proponents of environmental
sustainability, however, this argument rings hollow: As argued above,
barring extraordinary and unforeseen technological advances, resource-
intensive activities must occur *somewhere* in order for more-developed
countries to maintain the standard of living that they currently do. Thus,
environmental commentators harbor real concern that regulatory competition already may be creating barriers to sustainability and, in any
event, certainly may do so in the future as environmental regulation
compliance costs rise and create much stronger incentives for capital to
forum-shop.

[margin note: regulatory competition creates barriers to sustainability ✓]

Quite apart from this question of whether regulatory jurisdictions
can be expected to race to the bottom or to the top is a difference of viewpoint on the merits of racing at all. Economic reformers tend to favor
international regulatory harmonization on the ground that it lowers
transaction costs and facilitates trade. Moreover, they argue that highly
devolved, participatory decision making not only leads to cumbersome
heterogeneity in the standards governing commerce but also creates an
increased possibility of protectionist or alarmist lawmaking. Accordingly,
the perceived value of disciplining regulatory standards through high-
level scientific risk assessment and cost-benefit analysis lies both in rendering predictable the results of regulation and in tempering irrational
public demands, interest-group distortions, and other supposed dysfunc-
tions of participatory governance. From the environmental perspective,
in contrast, such centralized regulatory approaches are undesirable both
because they risk descent toward a lowest common denominator for environmental policy and because, more generally, they disrupt the goals of
regulatory experimentalism and sensitivity to local conditions. The latter
quality is thought to be especially desirable in light of the diversity and
complexity of local environmental conditions, cultural practices, and other
determinants of sustainability.

Sensitivity to local heterogeneity may be thwarted by international
trade liberalization in another sense. In addition to the desire to ensure

equitable and sustainable access to natural resources and ecosystem services, proponents of sustainable development typically also desire to ensure the survival of diverse human communities, languages, and traditions. The Rio Declaration on Environment and Development, for instance, observes that "[i]ndigenous people . . . and other local communities have a vital role in environmental management and development" and that states therefore must "recognize and duly support their identity, culture and interests."[26] This desire to preserve particular communities and traditions, however, may be undermined by the global integration of markets, particularly as trade expands in culturally inflected products such as media and consumer goods, and as relevant trade authorities drift from a legal standard of nondiscrimination to one more affirmatively in favor of market access for producers. Although it would be a mistake to equate human culture crudely with consumption patterns, it also would be a mistake to believe that the two are unrelated. Similarly, although the strongest critiques of advertising and other techniques of influence used by consumer product manufacturers are undoubtedly overstated, one would be unwise to dismiss altogether the possibility that billions of dollars of annual commercial messages help to shape desires, culture, and, at least indirectly, the environment.[27] Thus, to the extent that trade liberalization fosters not only enhanced efficiency in the satisfaction of existing preferences but also enhanced opportunities to participate in the evolution of new ones, the goals of trade liberalization and cultural heterogeneity may well be in conflict.

Finally, as Herman Daly has argued, liberalized trade may exacerbate the already considerable challenge of maintaining economies within environmentally sustainable limits "by making supplies of resources and [pollution] absorption capacities anywhere simultaneously available to demands everywhere."[28] To Daly, environmental sustainability already presents informational and governance challenges of a daunting magnitude on the national level; allowing importation of natural resources and exportation of wastes across the globe may mean that states face even duller incentives to monitor resource consumption, pollution loading, and other ecological implications of their activities. In turn, global market integration for natural resources may lead to a situation in which water and other basic means to life are subjected to competitive economic bidding with little or no public oversight. The greater the degree of privatization and commodification of natural resources and ecosystem services, the more powerful grows the case that their exchange should be governed by GATT disciplines and other rules of international trade, so that countries ultimately may be prevented from reserving freshwater or other vital "goods" for use by their own citizens. From this perspective, the ability of countries to ensure distributive equity and environmental sustainability in their activities at the national level—still the primary level at which effective, participatory governance exists— may be seriously hampered by the global integration of markets.

Because countries seem unlikely to abandon or reverse the trend toward global market integration, the natural response would seem to be that the global community should guide the global market through global regulation. As noted at the outset of this chapter, a primary thrust of environmentalism is precisely along these lines, aiming to reform laws and policies to better acknowledge global interdependence. One key site for such advocacy has been the WTO, where environmentalists have sought to establish the legitimacy of trade-based environmental regulatory efforts, such as labeling requirements, processing standards, and environmental tariffs, and have even proposed the establishment of broad environmental regulatory authority within the WTO, in view of its relative strength among international legal institutions. This demand for integrated analysis of economic activity and environmental conditions, however, stands in sharp contrast with the separationist tendencies of trade specialists, who see the legitimacy and effectiveness of the international trading regime as being dependent on a perception that its work is purely nonpolitical and technical in orientation—much as the demand for integrated analysis of efficiency and equity stands in sharp contrast with the desire of welfare economists to remain agnostic on issues of distribution and other overtly moral and political concerns.

As the remainder of this chapter demonstrates, these separationist tendencies also appear in recent game-theoretic treatments of international relations, which take the already atomized world of Westphalian sovereignty—the world in which, as Philip Allott notes, "humanity has chosen to regard its international world order as an 'unsocial world'"[29]— and reduce it further to a strategic competition among states for power and resources, with each actor being presumed simply to maximize its own interests, much as the individual actor within neoclassical economics is presumed to seek the optimal satisfaction of self-interest. Although these schools of thought advance beyond the EKC and other laissez-faire economic arguments by acknowledging an explicit role for international regulatory cooperation, they nevertheless depict such cooperation only as an unstable suspension of otherwise endemically rivalrous relations between states. Through this construction, the question of what *should* comprise a state's interests—of whether, for instance, altruistic regard for or moral obligation to other states should form a basis for its laws—is separated from law itself and is only discussed and negotiated, one must assume, in private forums that never enter into the view or grasp of the welfare-economic policy-making framework.[30] As will be seen, this reductionist view is ill suited even for maximizing its own narrow conception of state interest, given the myriad ways in which national well-being is dependent on the actions and decisions of other sovereigns. From the environmental perspective, an unsocialized international policy space is also an unsustainable one.

*

At the outset of this discussion, it should be noted that a long tradition exists within political theory of viewing dimly the prospects for enhancing justice and distributive equity across borders. In one sense, this dim perception is difficult to square with liberal convictions; as Paul Kahn has noted, "[i]f we are serious about our liberalism, we are led ever outward toward a global extension of transnational forms of government."[31] Accordingly, political theorists such as Bruce Ackerman[32] and Thomas Pogge[33] attempt to push liberal theory ever outward, both in ideal terms and in terms of the proper liberal stance on concrete policy issues such as international trade, immigration, and natural resource exploitation. Nevertheless, many other scholars follow Thomas Hobbes's realist approach and conclude that the conditions of justice—that is, the physical, social, and psychological conditions that are thought to be necessary before the project of justice can even get off the ground—simply do not attain between states. As one writer strongly put it, "[A]ll theories that advocate a global principle of socio-economic justice fail because the strong duties of economic or political justice do not arise amongst strangers in different states. Theories of global justice make a mistake about the domain of justice."[34] Even John Rawls, whose difference principle

provides one of modern political theory's most powerful egalitarian ideals, balked at the prospect of equality between states, arguing in *The Law
of Peoples* that the psychological and moral condition of deeply felt
reciprocity—which he took to be necessary to ground something like the
difference principle—exists only among citizens who share a common
history and a common self-understanding as members of a single political enterprise. Moreover, in response to those who would press the obvious gross disparities in living conditions throughout the world in service
of a demand for global justice, Rawls argued that states' histories are
determined not by unequal endowments of natural resources, but by the
successful or unsuccessful exercise of their own human capital and political culture, for which they are to be held responsible. Thus, apart
from minimal duties to assist other "peoples" whose dire living conditions prevent them from achieving a stable (even if nondemocratic and
illiberal) society, *The Law of Peoples* is not concerned with whether "the
well-being of the globally worst-off person can be improved."[35]

Seen in this light, economic analyses of global environmental issues
that deploy "objective" value-of-life criteria, such as per capita income
levels, provide a more egalitarian treatment of foreign citizens than
Rawls would require: *Some* value accorded to the well-being of other
states' residents is, after all, better than none. Such global cost-benefit
analyses are unusual, however, in that they disregard the distinctions
between political communities. Most cost-benefit analyses remain domestic in orientation, generally ignoring welfare impacts of domestic
policies abroad and, conversely, either ignoring or only awkwardly estimating domestic consequences of extrajurisdictional activities. In theory, of course, any state's effort to maximize domestic well-being should
include the level of altruistic regard that citizens actually have for foreigners. One analysis of foreign aid practices, for instance, determined
that individuals in the United States implicitly value the well-being of
poor citizens in developing countries at $\frac{1}{2,000}$ of the value of an American citizen.[36] Armed with this data, cost-benefit analysts could incorporate extraterritorial impacts of domestic environmental policies in a
manner that remains faithful to the welfare-maximization conception.
Nevertheless, the assumptions implicit in this approach—that policy
makers are only delegated power to represent domestic citizens and that

representation consists only of rote interest maximization—effectively disable the political community from reevaluating its regard for and responsibility to citizens in other states.

To bridge this divide between welfare calculation and extrajurisdictional features of policy problems, a growing number of scholars have advanced an intermediate welfarist approach to the understanding and determination of law and policy. Rather than ignore political boundaries altogether (as do global cost-benefit analyses) or ignore the world outside the instant political boundary (as do conventional domestic cost-benefit analyses), welfare analysts instead should encompass the choices and behaviors of other states by modeling and predicting them, much as individuals are modeled and predicted under the rational-actor conception. States, thus, should be understood as rational actors that "act instrumentally to achieve certain ends,"[37] seeking to maximize state interests through cooperative and competitive relations with other political entities. This approach has both positive and normative dimensions. From the positive perspective of understanding and predicting government behavior, state interest should be understood as tantamount to the aggregate well-being of a state's citizens (although inevitable imperfections of government representation, such as agency costs, poor information, and disloyalty, will mean that the state's policy choices do not perfectly track domestic well-being). Similarly, from the normative perspective of prescribing rules of international law, state interest should be equated with the aggregate well-being of the state's citizens, since as Eric Posner puts it, "[s]tates themselves are not moral agents," and "state interests are just constructs based on the interests and values of people living within states."[38]

Despite deeming states morally irrelevant, the intermediate welfarist approach of scholars like Posner does treat states as *politically* central. In theory, because states as collectivities are morally irrelevant, the ideal government should be a world government, one that passively and impartially optimizes laws for the well-being of the planet's entire human population. However, the presence of transaction costs, preference clusterings, economies of scale, and other real-world variables are thought to make the optimal size of government less than total; hence, according to the welfare-economic story, the Westphalian nation-state system, in which each "government chooses policies that maximize the welfare of its own citizens but ignores the welfare of citizens of other states,"[39] arises out of practical necessity. Citizens bind together to create states to represent them as "agents," not in the philosophical sense of a "thick" subject with identity and responsibility, but in the economic sense of a functional intermediary charged only with maximizing the interests of a principal. As Allott notes, this double existence—in which the state is virtually nonexistent as a domestic personality but is exclusively charged with international legal ordering—tends to reinforce the view that individuals can only be expected to bind together in a justice compact within, rather than across, existing communities: The state's double existence "requires each of us to be two people—with one set of moral judgments and social aspirations and legal expectations within our own national society, and another set . . . for everything that happens beyond the frontier of our national society."[40]

Through this observation, Allott suggests a significant complication for the welfare-economic account of the nation-state system: Rather than simply flowing from the preferences that individuals have with regard to the optimal size of their political community, the Westphalian system may actually be playing a role in constructing and perpetuating those preferences.[41] As with domestic environmental law and its constitutive potential, this possibility of endogeneity should slow our embrace of the welfare-economic perspective on international ordering, in which "international law emerges from and is sustained by nations acting rationally to maximize their interests . . . , given their perception of the interests of other states, and the distribution of power."[42] Such a perspective does not merely reflect but also partially determines our acculturation. Thus, it is appropriate to ask whether this bleak vision of international obligation is really one that we desire to hold up as a compass.[43] International law seen from the rationalist vantage point does not represent "real law," in the sense of embodying a set of values and expectations that the law's subjects accept as legitimate reflections of, or even constitutive elements of, their community. Accordingly, as Jack Goldsmith and Eric Posner put it, "nations have no moral obligation to comply with international law"; nor, given the conclusion of political theorists that conditions of justice do not attain between states, do "liberal democratic nations have [a] duty to engage in the strong cosmopolitan actions so often demanded of them."[44] Instead, states simply have a duty to deploy international law in a manner strategically calculated to maximize the well-being of their own citizens. Treaties and other international agreements, in this sense, do nothing more than "provide[] a focal point for coordination, and establish[] what counts as cooperation in a prisoner's dilemma."[45]

As anyone who has studied the prisoner's dilemma knows, the formal structure of the game allows little hope for cooperation, for mutual recognition and communication, or, more broadly, for social and moral growth. Similarly dim become the prospects for successful international cooperation from the rationalist perspective: A single "risk monster" country may threaten unpalatable consequences for global environmental problems irrespective of other countries' levels of precaution, just as one "utility monster" threatens unpalatable consequences on the individual level under interest-maximization theories.[46] Even widely embraced international agreements may not solve collective action dilemmas, given that, from the perspective of rational-choice theory, the likelihood of state shirking and consensus breakdown increases as the number of parties to a treaty expands.[47] Thus, the rational-choice-theory approach to international ordering reintroduces the collective action problem at the very moment that international law appears to be succeeding, counseling states to remain isolationist or to seek only bilateral or tightly regional arrangements, rather than to pursue a more visionary and inclusive form of global justice. The upshot of these theoretical claims is blunt: "Given the multiple conflicting interests of states on various issues, and the particular distribution of state power with respect to those issues, *many global problems are unsolvable.*"[48]

The influence of this way of perceiving international law and international relations was apparent during the U.S. Supreme Court's first engagement with global climate change. In *Massachusetts v. EPA,* a coalition of states, local governments, and civil society organizations challenged the EPA's failure to regulate greenhouse gas emissions under provisions of the CAA that mandate regulation of all motor vehicle emissions that "cause, or contribute to, air pollution which may reasonably be anticipated to endanger public health or welfare."[49] The Court's analysis of the claim focused centrally on the question of legal standing, which required the challengers to demonstrate that they had suffered a concrete and personal injury, that the injury was fairly traceable to the EPA's allegedly unlawful conduct, and that the injury was likely to be redressed by the requested relief.[50] The agency, in defense of its nonaction, argued that climate change presents "a unique collective action problem,"[51] and that unilateral action by the United States could easily be undone by increased emissions abroad, particularly from the rapidly developing and populous countries of China and India. In essence, the government invoked a quite familiar consequentialist argument—"It makes no difference whether or not I do it"[52]—that seeks to absolve agents from responsibility for overdetermined harms. With respect to individuals, the argument exploits features of the classical liberal framework for government action, in which private autonomy can be curtailed by the state only in order to prevent or remedy a harm that one actor clearly has imposed on another. With respect to states, the argument exploits the Westphalian international legal order, in which state sovereignty confers authority to engage in most any domestic act so long as the act does not give rise to a significant and identifiable transboundary harm. For environmental dilemmas, which multiply dramatically the number of actors implicated and render speculative or indirect the chains of causation involved, harms do not come home; they do not attach to any particular agent that satisfies the criteria of responsibility.

In addition to this consequentialist alibi, the government in *Massachusetts v. EPA* also offered an argument based on the rationalist theory of international relations. Specifically, the government argued that unilateral action by the United States could actually be counterproductive because it would "hamper the President's ability to persuade key developing countries to reduce greenhouse gas emissions."[53] On this account, because states can only be expected to behave self-interestedly in response to various sticks and carrots employed in international negotiation, abatement of domestic greenhouse gas emissions in advance of a multilateral agreement would forgo valuable bargaining power that is held currently by the United States due simply to the profligacy with which the country pollutes. At oral argument, conservative members of the Court appeared to embrace both lines of the government's instrumentalist reasoning. Justice Roberts observed that the challengers' position on redressability "assumes there isn't going to be a greater contribution of greenhouse gases from economic development in China and other places that's going to displace whatever marginal benefit you get here."[54] Justice Scalia similarly appeared to adopt the government's concern about prematurely binding the United States before multi-lateral policies were in place: "I presume the problem [the EPA has] in mind is that we have nothing to give in international negotiations. If we have done everything we can to reduce CO_2, you know, what deal do we make with foreign nations? What incentive do they have to go along with us?"[55] This sentiment was echoed by Justice Alito in a skeptical query to the challengers: "[W]hat is wrong with [EPA's] view that for the United States to proceed unilaterally would make things worse and therefore they're going to decline to regulate for that reason?"[56]

Despite the apparent influence of these arguments, the Supreme Court in *Massachusetts v. EPA* ultimately rejected the government's position, with five justices agreeing that the elements of legal standing had been successfully demonstrated and that the language of the CAA required the EPA to ascertain whether greenhouse gas emissions endanger the public. Still, the sense of an unbridgeable strategic impasse continued to hamper climate action in Washington. As Congress prepared to consider for the first time significant greenhouse gas emissions legislation, the ranking Republican member of the Senate Energy Committee, Pete Domenici (N.Mex.), stated that he would "kill" any climate change bill that did not require China and India to meet the same standards as the United States. His spokesperson stated simply, "Whatever we do, he wants China and India to do it, too." Similarly, after *Massachusetts v. EPA* was decided, President Bush stated that "whatever we do must be in concert with what happens internationally." In particular, he emphasized that "[u]nless there is an accord with China, China will produce greenhouse gas emissions that offset anything we do in a brief period of time."[57] Oddly, even as the domestic political situation in the United States began to become more favorably disposed toward meaningful climate change legislation, the country continued to display aggressive opposition to internationally binding commitments. When Germany, as host of the 2007 G8 Summit, proposed a temperature-based approach that would obligate nations to reduce emissions sufficiently to avoid increasing global temperatures no more than 2 degrees Celsius (a target that experts estimated would require a reduction in emissions levels to 50 percent below 1990 levels by 2050), the United States responded that it was "fundamentally opposed" to such legislative proposals and that they "crossed multiple 'red lines' in terms of what we simply cannot agree to."[58] A particularly revealing moment came when the European Union proposed integrating airline emissions into its regulatory scheme, including a requirement that U.S. airlines landing in Europe purchase greenhouse gas emissions credits. In response, one U.S. policy maker offered a vehement assertion of national independence, one that seemed to assert territorial control over European landing strips: "This is our sovereignty. We'll deal with it."[59]

This view of foreign relations as an essentially hostile and noncommunicative game of chicken also became apparent in President Bush's statement that "each country needs to recognize that we must reduce our greenhouse gases . . . to come up with an effective strategy that, hopefully, when added together . . . leads to a real reduction."[60] Even from the perspective of unbridled state self-interest, this conception must fail, for it presupposes a vision of state autonomy and self-sufficiency that can no longer be maintained: Under circumstances of deep environmental interconnectivity, the state or other relevant political community cannot be content to go it alone or to simply "hope" that other states' policies will lead to a sustainable planet "when added together." The political community instead must perceive itself as a subject that stands in relations of responsibility and dependency with other significant actors on the geopolitical stage. It must, in other words, perceive itself as a collective subject that governs its conduct according to carefully reasoned values and aims within a context of constrained autonomy, while respecting other collective subjects by appealing to *their* ability to reason and decide within that same unavoidably constrained and tragic context. This is, in part, what Allott means by the necessity to socialize the international legal order. Often, one can discern something of this "thicker" social world in the rhetoric that political actors use to describe the history, identity, and responsibility of the collectivities they represent—rhetoric that suggests a much more nuanced collective personality than the stripped-down interest maximizer of the welfarist approach. For instance, when the British government adopted landmark greenhouse gas legislation in early 2007, then–Prime Minister Tony Blair hailed the moment as a "historic day" that "sets an example for the rest of the world."[61] Blair's unlikely political compatriot, California Governor Arnold Schwarzenegger, recognized the import of Britain's action: "It is very clear the Prime Minister has been a great inspiration to many, many countries all over the world. . . . I think he is a pioneer, because he has had the guts to sign the Kyoto treaty and to show to the world that you can protect the environment and the economy at the same time."[62]

Further evidence of this socialized international space came from the prime minister of Japan, who, after observing the division between the European Union and the United States on climate change, expressed a sense of national obligation to help lead climate negotiations: "I feel that it is our important responsibility."[63] Observers also have remarked that Japan feels a special sense of commitment to the much-beleaguered Kyoto Protocol, given its status as host country for the document's negotiation.[64] If accurate, this observation is revealing: Only a polity with a shared sense of collective identity—one that extends beyond merely a practical agreement to adopt institutions and procedures for maximizing individual interests—would perceive its legacy as being bound up with the success of a document that happens to bear one of its cities' names. More generally, the fact that most of the industrialized world pressed on with the Kyoto Protocol, despite the underinclusiveness of the document and the unwillingness of the United States to participate in the regime, suggests that, to those countries, expected outcomes do not fully determine the normativity of state action. After all, why pursue domestic restrictions if they will have no consequential impact—if the United States, China, or India could each individually obliterate the beneficial effect of greenhouse gas restrictions in Europe and Japan? The answer, perhaps, is that the basis of climate policy in those countries extends beyond interest optimization to include a conviction that political communities owe each other certain responsibilities that must be discharged with care and caution in light of the deep uncertainty that accompanies global environmental disruption. Notwithstanding the ready availability of self-interested and other *realpolitik* explanations,[65] it is worth retaining the notion that those countries may actually hope to inspire and lead others to a new shared perspective on climate change, believing that perceptions of state interest not only drive international relations but also can, at least in part, be determined by them.

At oral argument and in his dissenting opinion in *Massachusetts v. EPA*, Chief Justice Roberts dismissed out of hand this vision of a socialized international policy space, sharply rejecting the challengers' argument that unilateral action by the United States on climate change might inspire other states to take similar steps.[66] In essence, Roberts sought to enshrine into legal standing doctrine the rational-choice-theory approach to international relations. He argued that "the domestic emissions at issue here may become an increasingly marginal portion of global emissions, and any decreases produced by petitioners' desired standards are likely to be overwhelmed many times over by emissions increases elsewhere in the world."[67] Citing an earlier standing opinion of Justice Kennedy,[68] he argued that the challengers were required to affirmatively demonstrate that regulatory action by the EPA would prompt other countries to follow suit. Because Roberts took as given a foreign context in which emissions will continue to grow and to undo whatever unilateral action U.S. regulators take, he dismissed as a "sleight-of-hand"[69] the majority's argument that agencies can properly redress threatened harms by "whittl[ing] away at them over time, refining their preferred approach as circumstances change and as they develop a more-nuanced understanding of how best to proceed."[70] To Roberts, the possibility that domestic U.S. steps toward greenhouse gas emissions abatement would alter the international policy arena seemed simply unimaginable, a skepticism that ignored the earlier example of ozone-depleting substances in which American leadership had encouraged precisely that effect.[71]

The chief justice's restrictive views on redressability in the global environmental context deserve serious scrutiny: If the test for standing were made to depend so heavily on consequentialist efficacy in this manner, even domestic environmental laws would not fare well, for many of the goals of those laws face serious disruption from extraterritorial action. Increasingly apparent, for instance, is the internationally dependent nature not just of ozone depletion and climate change but of all air quality regulation. To be sure, the international community has long recognized the problem of transboundary air pollution between certain states, such as the United States and Canada or the members of the European Union.[72] In the past decade, however, emerging scientific evidence of transboundary air emissions from East Asia and their impact on air quality in North America has highlighted the fact that air pollution regulation is much more deeply affected by extraterritorial activities than previously appreciated. For several air pollutants regulated under the CAA, including ground-level ozone precursors, particulate matter, carbon monoxide, and mercury, the ability of state and federal authorities in the United States to achieve air quality standards is becoming increasingly frustrated by periodic transpacific pollutant plumes and by more-general East Asian contributions to background-pollutant concentration levels. After 2020, for example, when Asian nitrogen oxide emissions are expected to quadruple from 1990 levels, transpacific ozone precursor plumes could increase springtime ozone concentrations in California by 40 parts per billion by volume (ppbv).[73] To put this number in perspective, as of early 2010 the national ambient air-quality standard for surface ozone concentration was 75–80 ppbv, averaged over eight hours.[74] Already scientists estimate that increased Asian emissions of the ozone precursors NO_x have caused a 30 percent increase (10 ppbv) in background-ozone concentrations along the western United States since the mid-1980s.[75]

These transboundary effects place U.S. states in a quandary: In developing their air quality implementation plans under the CAA, states must include within their planning anticipated emissions from transboundary sources, such as actors in other U.S. states or foreign countries,[76] yet they themselves have little authority to control or influence those extrajurisdictional sources. As Justice Stevens observed in *Massachusetts v. EPA* with respect to the problem of climate change, "Massachusetts cannot invade Rhode Island to force reductions in greenhouse gas emissions, [and] it cannot negotiate an emissions treaty with China or India."[77] Thus, states under the CAA find themselves in a position much like the one that the welfare-economic framework foists upon policy makers, treating extrajurisdictional sources as an empirical given rather than as a subject of recognition, reason, and responsibility. As more information is discovered about the environmental, social, and economic costs of Asian emissions, the need for U.S. regulators to seek cooperative relations with China and other countries will become increasingly evident, not just with respect to obviously global issues such as climate change but also with respect to areas of traditionally domestic concern. In that sense, the welfare-economic framework for government policy making will become increasingly plagued by an awareness that other countries hold a subjectivity that cannot be empiricized, that the integrity of the Westphalian order has been compromised by ubiquitous externalities that inevitably encroach somewhere internally, and that the power of state policy, when made to depend on rationalist interest maximization under these conditions, evaporates into choked air.

*

Those entities or locations that appear to us as discontinuous—those "territories," "states," and "persons" that become the conceptual inhabitants of our legal geography—remain deeply embedded within systems that resist ready dissection, comprehension, and control. We are presented not with discrete natural environments connected only by certain global common substrates, nor with discrete political communities connected only by certain channels of international commerce and environmental impact. Rather, we are presented with a "complicated tissue of events," both biophysical and sociolegal, in which even conventionally domestic environmental problems must be viewed as global in scope and in which politics and law accordingly must adapt to the challenge of ineradicable interdependence among states. Rather than promote such adaptation, the rise of the risk-assessment-and-cost-benefit-analysis paradigm has made it more difficult to perceive and refine the United States' national environmental identity. When other states are forced to appear within policy assessments merely as inputs to empirical calculation, environmental problems appear intractable, insoluble, wicked. In such noncommunicative contexts, the only available brand of rationality appears to be that of purely strategic, self-interested behavior, a logic that narrows the apparent scope of possible resolution to a thin band of fortuitous and ever-contingent mutual accords. In turn, this narrow instrumentalism leads naturally to a sense of hopelessness and to the conclusion, expressed prominently by dissenting jurists in *Massachusetts v. EPA*, that domestic environmental law should not be interpreted to encompass harms caused by climate change and other global environmental processes, given that unilateral action by the United States could at best mitigate only a small portion of those harms.

This seemingly powerful logic casts a greater pall than the dissenters acknowledged. The problem of transpacific air pollutants suggests that the *Massachusetts v. EPA* dissenters' reasoning not only would block litigants from challenging the EPA's failure to regulate greenhouse gas emissions; it also would enable the abandonment of a great deal of the existing CAA's ambient air quality program, since that program no longer can be considered independently of the extrajurisdictional decisions that will, in substantial part, determine its efficacy. Thus, even from the purely self-interested vantage point assumed by welfarist approaches, law's geography must become more expansive, more permeable, more deeply interpenetrated. Yet the multilateral moment required to achieve progress on air quality—and on numerous other environmental problems—seems unlikely to arrive amidst the predominance of a policy framework that fails to promote a level of collective self-awareness commensurate with the demands of international dialogue.

Law's geography must become more expansive

As the next chapter explains, a similar theoretical problem plagues matters of intergenerational justice. As with the realist position in international affairs, many political theorists take the circumstances between generations to be incapable of supporting justice relations. Because future generations are always yet to come, they are structurally vulnerable to exploitation by the present generation, most notably through the all-too-tempting device of temporally externalized costs. In this view, generations are believed to be essentially presentist in orientation, just as states are taken to be purely self-interested.[78] Even those generations that might desire to depart from such presentism will be deterred by the fact that their efforts to promote the interests of future generations—for instance, by setting aside financial resources or undeveloped land—will always be vulnerable to self-interested raiding by intervening generations. Accordingly, the relation between generations is thought to resemble that between states, with a similar sense of silence and hopelessness.

Nevertheless, as the next chapter argues, even generations can be understood to be interdependent and socialized, just as states can be. Despite time's unidirectional arrow, it is not the case that "there is no mechanism by which [future generations] can transfer resources to us,"[79] or, conversely, cause us harm. The temporal relation between generations is characterized by asymmetry and by a tyranny of the present, but there is at least one desideratum that future generations *can* offer in exchange for our conscientious stewardship of their interests: their respect. Likewise, future generations also can harm us by renarrating our history and by altering the meaning of our identity in line with their assessment of how well or how poorly we did by them.[80] Thus, just as we should move beyond narrow rationalist accounts of international relations, we also should move beyond an account of intergenerational relations in which the well-being of future generations depends entirely on our generosity. Again, the world wants for justice, not charity.

CHAPTER SIX

Other Generations

DEPICTING STATES WITH STRONG subjectivity cuts against a dominant progressive thought in international law these days, which is that the state should be de-emphasized as a legal personality on the global stage, or at least that nonstate actors such as advocacy groups, indigenous communities, business alliances, and other entities should be given greater standing and prominence.[1] The argument of the previous chapter, however, does not deny the need for a more inclusive policy-making discourse at the global level, nor does it contend that the state is an unproblematic vehicle for recognizing and redressing policy problems with global environmental dimensions. Instead, the argument simply contends that successful global environmental governance requires, as a minimal sufficiency condition, something more than rationalist welfare analysis can provide. An essential premise of that framework is that collective choice should passively and impartially trace the results of an individualized welfare calculus; government policies thus are not attached to an identifiable agent who bears responsibility for their content or effect. Because the state remains the primary geopolitical unit at the dawn of the twenty-first century,[2] the personality of states must be seen to encompass more than merely a set of instructions regarding risk assessment and welfare maximization; similarly, the social world of states must be seen to entail more than simply a playing field for competitive maneuvering around fixed interests or goals. By accepting a notion of collective responsibility, such as the one implicit within the precautionary principle, the state can perceive itself as an independently significant actor on the geopolitical stage, one that stands in relations of dependency and obligation with respect to other sovereigns. Such a relational viewpoint, in turn, can highlight the need for international leadership and cooperation.[3]

As noted in chapter 2, this self-consciously collective or organicist conception of governance is in tension with the individualistic thrust of liberal political theory. Although liberalism comes in many stripes, at the core of liberal theories tends to be a belief that the individual is, if not ontologically prior to social groups and orderings, then at least normatively privileged in the sense of providing the proper vantage point from which to consider government obligations to protect and provide. With respect to future generations, a host of complications arise from this vantage point due to the inescapable fact that future individuals do not yet exist. Like children, future generations are part of the "Achilles' heel of liberalism"[4]—that vulnerable location for agents who are imperfectly situated to assert their rights or interests in the manner that liberalism often demands of them. So long as we regard the hope of self-determination to be foundational to our political philosophy—which no doubt we should do in some fashion vis-à-vis the presently living—then members of future generations will always appear to us as other, as incapable of fully entering into our social partnership.[5]

And yet our environmental laws ask us to do precisely that, to bring future generations cognitively and programmatically into our public decision making. When the U.S. Congress in late 1969 listed six specific obligations for the federal government as part of its visionary introduction to the NEPA statute, it listed first the need to "fulfill the responsibilities of each generation as trustee of the environment for succeeding generations."[6] Congress's phrasing recalls the Great Law of Peace of the Iroquois Nation, which similarly observes that tribal decision makers must "[l]ook and listen for the welfare of the whole people and have always in view not only the present but also the coming generations." In order to implement this duty of regard for the unborn, the Iroquois Constitution offers a simple and now much celebrated decision-making heuristic: "In our every deliberation we must consider the impact of our

Achilles' heel of liberalism

7 generations heuristic

decisions on the next seven generations."[7] Today, the centrality of future generations to discussions of environmental law and governance is reinforced by the concept of[sustainable development] Like the precautionary principle, this concept has attracted adherents throughout the international environmental community, despite staunch claims by critics that the notion is vague and indefensible.[8] Its affective pull stems perhaps from its promise to bring future generations before us, for why else should we sustain the integrity of the environment, if not for them?

Something about the ethic of futurity does resonate deeply. It seems to promise that entire worldviews and political philosophies can be reduced to bumper-sticker mantras: "We do not inherit the Earth from our ancestors: We borrow it from our children." Such popular sentiments differ little from those of an earlier pamphleteer, Thomas Paine, who in his seminal statement of the Enlightenment ideals of liberty and equality proclaimed that "[t]he illuminating and divine principle of the equal rights of man . . . relates, not only to the living individuals, but to generations of men succeeding each other."[9] More than a century later, Theodore Roosevelt deftly upended utilitarian reasoning—which might have counseled widespread economic development of forests and other wilderness areas—by observing that "[t]he 'greatest good for the greatest number' applies to the number within the womb of time, compared to which those now alive form but an insignificant fraction." Accordingly, in Roosevelt's view, few present economic benefits could be substantial enough to outweigh the stream of recreational, aesthetic, and spiritual benefits that would flow to generations of Americans if vital wilderness areas are preserved: "[O]ur duty to the whole, including the unborn generations, bids us restrain an unprincipled present-day minority from wasting the heritage of these unborn generations."[10]

"the number within the womb of time"

Importantly, to Roosevelt, this duty of heritage preservation and stewardship did not imply a tyranny by the future over the present. On the contrary, he saw it as reflecting an expanded democracy, one in which the needs and interests of future generations are represented on a par with those of the present: "The movement for the conservation of wild life and the larger movement for the conservation of all our natural resources are essentially democratic in spirit, purpose, and method."[11] Decades later, prominent political theorists joined Roosevelt in holding that there is no essential tension between liberal democracy and inter-generational justice. Most prominently, when elaborating *A Theory of Justice*, John Rawls set out a "just savings" principle of intergenerational responsibility, according to which each generation must "put aside . . . a suitable amount of real capital" in order to fashion and preserve "just institutions and the fair value of liberty."[12] Rather than apply his famous difference principle of distributive justice intergenerationally, Rawls asked only that each generation contribute to an accumulation of capital that "make[s] possible the conditions needed to establish and to preserve a just basic structure over time,"[13] after which the demands of intergen-erational justice would be satisfied and the net-capital-accumulation obligation would fall to zero. Moreover, unlike Roosevelt's conservation-ist duty, which requires the preservation of heritage areas and other *specific* natural resources, Rawls's just-savings principle merely requires intergenerational transfer of an undifferentiated accumulation of "real capital," which Rawls construed in broadly economic terms as having to do with the material base of social institutions.

Rawls
contra
Roosevelt
✓

Still, much as Roosevelt believed that conservation of natural re-sources flows smoothly and without contradiction from an expanded de-mocracy, Rawls believed that the just-savings principle can be satisfied in a fashion that both ensures the well-being of future generations and avoids undue government definition or restriction of the life courses of its citizens: "While all generations are to do their part in reaching the just state of things beyond which no further saving is required, this state is not to be thought of as that alone which gives meaning and purpose to the whole process. To the contrary, all generations have their appropriate aims."[14] Nor is Rawls the only liberal luminary to take this optimistic view. Ronald Dworkin similarly suggests that liberals can properly favor government-sponsored environmental conservation out of a desire to preserve the ability of future generations to revere and recreate in natu-ral areas.[15] In his view, conservation does not violate the liberal stance against government arbitration among competing conceptions of the good, since a conservationist government merely seeks to preserve the option for future generations to value nature, if they choose to do so. In his classic work *Social Justice in the Liberal State,* Bruce Ackerman simi-larly notes that the obligation of liberal trusteeship should seek to preserve an "overall pattern of objects [that] reflect[s] a sense of the rich variety of possibilities that the future may have in store."[16] While he acknowledges that under some circumstances irreplaceable resources like the Grand Canyon might be preserved consistent with liberal neutrality, he tends to agree with Rawls that bequests to the future can generally be cashed out in terms of fungible economic resources.[17]

Option-preserving positions

These option-preserving positions have an undeniable appeal. Still, they beg the question: What if the conservation of ecological options for future generations *does* preclude ways of living desired by the present generation? What if we have passed the period in which increased productivity, technological development, and the opening of new frontiers can be depended upon to offer an [escape valve] from the reality of ecological finitude? What if, instead, we now are faced directly with the

No escape valve

environmental claims of the future? Political theorists often seem to crave Locke's ecology, where "enough, and as good" natural resources are believed to exist to avoid the necessity of collectively determining resource uses.[18] Indeed, sometimes the condition of only moderate, as opposed to strict, resource scarcity, is taken to be a precondition for the very functioning of liberal rules of justice.[19] Depending on its articulated scope, however, the futurity obligation may well place serious limitations on a liberal government's ability to afford presently living individuals ample space for private activity and self-determination. Particularly when affluent consumers—and many of them—regard their conception of the good as being dependent on the acquisition of more and better material goods, liberalism may well be in conflict with the needs and interests of future generations.

Coral reefs example

To give just one concrete illustration, suppose we were to adopt a goal of preserving living coral reefs for possible enjoyment and appreciation by future generations. To achieve this goal, we would need to legislate immediate and drastic changes to the lifestyles of presently living individuals in many nations, given that the energy, transportation, agriculture, and land use patterns associated with those lifestyles currently threaten the survival of all living coral on the planet.[20] Thus, we would face a choice between, on the one hand, altering the life courses of the currently living or, on the other hand, altering the environment of future generations in a way that, in turn, will restrict their possible life courses.[21] A laissez-faire economic growth policy does not avoid this dilemma but instead makes the implicit determination that admiration of living coral reefs need not be a part of future generations' tapestry of options.

*

Notwithstanding its resonance, the standard conception of harmony between democracy and conservation, between liberal individualism and intergenerational environmental justice, may ring untrue: Not only does intergenerational environmental justice seem to presuppose the existence of some collective community of interest beyond the mere agglomeration of individual interests—for how else is one to conceptualize obligations between generations?—but it also seems to necessitate the imposition of resource policies that are premised on implicit or explicit assumptions about the nature of the life well lived. Indeed, this necessity of deliberate government endorsement or condemnation of particular technologies, preferences, and values already seems to have been acknowledged by most of the international environmental policy-making community: In language that offers a remarkably direct attack on the premises of market liberalism, the Plan of Implementation of the 2002 World Summit on Sustainable Development, in Johannesburg, identifies "changing unsustainable patterns of production and consumption" as an "overarching objective[] of, and essential requirement[] for, sustainable development.[22]

Ten years earlier, in preparation for the United Nations Conference on Environment and Development, in Rio de Janeiro, U.S. representatives infamously warned against precisely this line of thought by intoning, "The American way of life is not negotiable." One of the reasons that U.S. representatives in Rio could adopt, without obvious embarrassment, such an absolutist and seemingly self-serving stance is that they came armed with a theoretical edifice that claims to refute the need to engage in any such negotiation, an edifice that purports to show that the "American way of life" is not only sustainable but optimal for all generations to come. Specifically, the U.S. representatives came armed with the teachings of neoclassic growth theory and welfare-economic-policy analysis, both of which seem to have resolved the problem of intergenerational resource allocation in a manner that avoids the need to prioritize natural resource uses across time.[23] Importantly, these schools of economic thought claim to have accommodated intergenerational concerns in a manner that requires no serious adjustment to prevailing theoretical models or accompanying policy prescriptions. This is a striking claim, for moral and political philosophers regard the topic of intergenerational relations to be among the most vexed and demanding in all of human thought.[24] How could it be, then, that the problem of intergenerational relations admits of such a tidy resolution?

To be clear, this is not to say that economists believe that unregulated market activity will vouchsafe the environment on behalf of future generations, for they typically recognize that environmental problems do necessitate public regulatory efforts, such as the development and enforcement of tradable pollution permits or some more-direct control of responsible actors through regulation or taxation.[25] Nevertheless, proponents of the economic approach typically shape their policy recommendations in light of theoretical models that depict a market-based society that is somehow free of pollution externalities and other failures of private ordering. This reification occurs on the macroeconomic level in the form of [optimal growth theory,] which, at least in one prevailing form, aims to depict the pattern of consumption that would occur across time within an idealized society in which all individuals discount well-being at a uniform specified rate, all markets are both competitive and free of externalities, and a complete set of futures markets exists.[26] It occurs on the microeconomic level in the form of cost-benefit analysis, which, as earlier chapters explained, seeks to evaluate the consequences to well-being that flow from particular policy proposals through marginal cost-and-benefit assessment.

The attraction of these models is their implicit suggestion that so long as society "gets the prices right," environmental sustainability can be achieved through market dynamics alone, without need for collective determination regarding resource use and conservation policies. Still, care must be taken to distinguish between intergenerational efficiency and intergenerational equity under this construct. Even in the idealized macroeconomic model, the optimal consumption path does not guarantee attainment of any particular level of consumption over any particular time period. Put differently, although the "standard policy remedies for improving economic efficiency—like establishing property rights, addressing externalities and so forth"—may help to maximize the net present value of aggregate intergenerational utility, they "do not guarantee sustainability."[27] Instead, in order to constitute truly sustainable development, society must supplement the discounted welfare-maximization criterion with an additional requirement that discounted welfare not be permitted to decrease over time or, at least, that it not be allowed to fall below some baseline of minimally acceptable well-being.[28] As discussed below, one means of doing so—though not one typically accepted by economists—is to guarantee the provision of certain inalienable ecological goods and services.

Sustainability requires discounted welfare to not fall over time

The criterion of intergenerational efficiency can be thought of as conceptually analogous to the criterion of Kaldor-Hicks efficiency in the intragenerational context. Just as Kaldor-Hicks-efficient policies could, in theory, become both efficient and equitable (because the winners from the policies could compensate the losers and still come out ahead), intergenerationally efficient consumption paths could, again in theory, become both efficient and sustainable (because the winning generations could transfer enough resources to the losing generations to ensure a nondeclining welfare path).[29] Accordingly, just as Kaldor-Hicks efficiency often is known as potential Pareto efficiency, environmental economists suggest that intergenerational efficiency should be thought of as "potential sustainability."[30] By offering this conceptual clarification, welfare economists aim to establish separate domains of competency within policy analysis: Supporters of the economic paradigm will be able to focus their policy recommendations on the maximization of intergenerational efficiency, while advocates of environmental sustainability will be forced to rely on the "political process" of the tax-and-transfer system to distribute resources among present and future generations.[31]

Just as chapter 4 revealed the efficiency-equity construction to be a troubled one, the remainder of this chapter dissects the efficiency-sustainability construction, focusing particularly on its essential analytical fulcrum, the practice of exponential discounting. Within both optimal growth theory and welfare-economic-policy analysis, the problem of measuring well-being across time is addressed through the use of a discount rate, which attempts to "normalize" values that occur in different time periods. According to this procedure, an increment of value today is treated as more valuable than the same-sized increment tomorrow. For instance, at a 5 percent annual discount rate, 105 units of well-being in year two would be worth only 100 units today, in year one. Put more dramatically, at the same discount rate, "one statistical death next year counts for more than a billion deaths in four hundred years."[32] Whatever the wisdom of such a practice for intragenerational policy-making,[33] it raises distinct questions in the intergenerational context, where discounting over lengthy time horizons has the practical effect of diminishing dramatically the significance of policy effects that occur in the future. Indeed, the impact of discounting future costs and benefits to a present value tends to swamp all other variables within economic analyses of intergenerational challenges, such as climate change.[34] Thus, arguments in support of the practice of discounting merit close attention. Unless a normatively attractive case for discounting can be uncovered, the obvious unattractive implications of the practice seem to counsel strongly against it.

Discounting
swamps other
variables... ✓
↓
How?

*

Many of the arguments that have been offered in defense of discounting in the intergenerational policy context can be dismissed summarily. For instance, the possibility that future generations may not come into existence due to extinction—which sometimes is used to provide an analytical basis for discounting—offers little comfort in contexts where the likelihood that future generations will survive is itself a function of policy choices that are being made through discounted welfare analysis. In such contexts, analysts who promote discounting seem rather like the juvenile criminal defendant in an old joke who, having been accused of murdering his parents, pleaded for the court's mercy by proclaiming, "I'm an orphan." Similarly, the absurd or paradoxical results that are said to flow from a refusal to discount fail as defenses of discounting because they assume prior adherence by society to a mathematical optimization procedure, when in fact the very question being posed is whether intergenerational decision making is best managed through that type of social choice mechanism. Analysts sometimes argue, for instance, that future benefits should be discounted at some positive rate, because otherwise present generations operating under the welfare-maximization mandate will be forced to perpetually forgo present benefits in order to achieve the same level of benefits later at a lower present cost. The problem with this argument is that it merely demonstrates that the present generation might desire to avoid unduly sacrificing for future generations; it does not explain why discounting is the proper method for implementing such a desire.

Also problematic are justifications for discounting that are premised on the observed rate-of-time preference. The rate-of-time preference is thought to be an appropriate policy device because it can be found, as a descriptive matter, in the behavior of individuals. That is, currently living individuals seem to prefer present consumption to future consumption at some empirically identifiable rate, and thus analysts assume that a similar degree of impatience should be applied to the intertemporal allocation of consumption between generations.[35] In this context, however, the lens of methodological individualism leads to confusion, most obviously through the uncritical assumption that the observed impatience of individual members of the present generation should govern the consumption opportunities left available to future generations. If we truly aim to maximize individuals' actual preferences, then some more-deliberate attempt to predict and understand the rate of time preference of other generations is required. Moreover, by taking revealed preference as the exclusive standard for welfare analysis, the analyst fails to acknowledge that a traditional, and often quite popular, role of democratic government has been precisely to counteract the influence of consumer impatience. Our citizen selves frequently support policies that tie the hands of our consumer selves, such as through forced retirement savings or prohibitions on the advertisement of harmful goods.[36] This impatience-mitigating or hands-tying function may be all the more important when the interests of future generations are at stake, given that members of those generations suffer an even greater tyranny of the present than do the currently living. Proponents of the economic approach sometimes respond to such contentions by making plain their political commitments: "[O]verriding market prices on ethical grounds[]opens the door to irreconcilable inconsistencies. If ethical arguments, rather than the revealed preferences of citizens, form the rationale for a low discount rate, cannot ethical arguments be applied to other questions?"[37] This objection misfires: In truth, a need for "ethical arguments" and an "overriding" of preferences is inevitable no matter how the present generation proceeds, since the rate-of-time preference of future generations cannot be observed.

Even if future generations' rate-of-time preference somehow could be reliably predicted, a logical error still would infect the discounting practice, given that welfare maximization takes the relevant period of discounting to begin at the moment of policy calculation, before the lives of future generations have even begun. Again, if one is truly following the rationale of respecting individuals' preferences, then discounting of future costs and benefits for time impatience should begin and extend only for the time period that the individuals affected by policy proposals are actually alive, experiencing impatience. The welfare maximizer instead adopts a constant rate and an uninterrupted period of discounting, one that is detached from any actual individual lives-in-being, either now or in the future. The maximizer adopts this counterintuitive approach because he has entertained a subtle, unacknowledged shift from the aggregation of individual welfare impacts to the agglomeration of welfare impacts *as such*, in a disembodied form. The maximizer essentially treats all members of future generations as if they were a single infinitely lived interest holder. It is difficult to imagine, though, why future generations would consent to being treated in this manner—as if their lives were already in decline at the very moment that they come into being—when they themselves did not enjoy the period of youthful extravagance that was gained at their expense—by us. No matter how it is spun, discounting for rate-of-time preference means that future well-being is weighted less than well-being today, for no reason other than that it occurs later in time. Such a practice therefore stands in obvious tension with the principle of equal, or at least equitable, regard for future generations.[38]

Normative justifications for discounting that are premised on the significance of opportunity costs have a somewhat more compelling basis than the rate-of-time-preference argument. From this perspective, discounting future costs and benefits according to the rate of return available for alternative uses of public funds is necessary to ensure that policies will leave future generations with an endowment of resources that has taken advantage of the best available investment opportunities. Defenders of discounting argue that this practice serves future generations by ensuring that the stock of resources eventually bequeathed to them will have grown to its largest attainable size. In their view, society best serves the interests of future generations not by transferring any particular portfolio of assets but by maximizing the option set available to such generations, which in turn counsels nurturing the overall stock of useful capital by allowing resources to be devoted to their most valued uses. From this perspective, our present patterns of production and consumption, which were singled out for condemnation by the Johannesburg Plan of Implementation, appear instead to represent the desirable creative destruction of market economies: "[J]ust as earlier generations invested in capital goods, research and education to bequeath to current generations the ability to achieve high levels of consumption, current generations are making the investments that are necessary to assure higher real living standards in the future, despite stresses on the natural resource base."[39]

Nevertheless, the need to attend to opportunity costs does not by itself suffice to establish that discounting is the appropriate mechanism for doing so. Upon close inspection, the opportunity-costs argument turns out to be, like the rate-of-time-preference argument, riddled with complications and paradoxes. Most notably, in order to sustain what would otherwise amount to a gross conflation of potential and actual Pareto improvements, the opportunity-cost justification of discounting must assume that sufficient intergenerational resource transfers actually will be undertaken to ensure that future generations are left no worse off, even after their interests have been exponentially discounted in our policy making. As noted above, a vital difference exists between actual and potential sustainability in the economist's vernacular. Even if a society maximizes the value of its investments through the disciplinary effect of discounting, it still may need to engage in deliberate intergenerational resource transfers to ensure equity across generations. Such a task would require the welfare economist to roll up his sleeves and consider the details of a duty of intergenerational trusteeship, along the lines of Ackerman's approach within liberal political theory or of Herman Daly's within ecological economics.[40] Instead, the form and manner of intergenerational transfers is generally left unstated in the economic literature or is relegated to the same "political process" dustbin used to address matters of intragenerational equity.

Worse still, some analysts deem compensatory transfers entirely unnecessary, given the legacy of economic growth and technological progress that they are confident we will bequeath to future generations so long as public regulators simply stay out of the way. After all, they argue, recent human history suggests that each generation will be progressively better off than the previous one simply on account of the accumulation of knowledge and resources that occurs through unregulated market activity.[41] Like the extreme technological pessimism of some environmentalists, this "trickle forward" defense of discounting is more of a mood than an argument. Even viewed charitably, the defense misfires for, as Geoffrey Heal points out, it "is not an intertemporal judgment. It is an interpersonal one."[42] That is, the defense seeks to justify discounting impacts on future generations, not because they will happen in the future, but because they will happen to individuals who are expected to be better off than those who presently exist. This mistake is similar to the one discussed above, in which the cost-benefit analyst confuses discounting for time with discounting to avoid undue sacrifice. As a matter of equity, we may agree with both contentions, but they require different implementation devices than crude numerical discounting. Ironically, if we use standard economic methods for valuing life, then an anticipated rise in future incomes actually suggests that future lives are *more* worth saving than present lives, not less.[43] Thus, rather than discount future benefits of environmental protection because future generations will be better off, we should magnify them. The oddness of this result suggests that the normative justification for discounting has been insufficiently established. (It should also suggest that we are doing too little at present to adjust benefits that accrue to better-off and worse-off individuals within the *current* generation.)

Perhaps in light of these complications, many defenders of discounting stress that intergenerational distributive equity must be addressed through deliberate resource transfers and not left blindly to market processes. To give this position practical significance, however, governments must develop some more or less comprehensive system of public accounting to ensure that the resource base actually is being expanded (or at least preserved) for the benefit of future generations. At present, the danger is too great that consumption may be confused for investment, that environmental externalities may be inadequately accounted for, and that many important natural resources and ecosystem services may be absent from national ledgers altogether. Moreover, if the system of environmental accounting revealed that the total capital stock was not being preserved adequately for the benefit of future generations—as many expect it would—then some socially controlled mechanism of intergenerational capital transfer would be required in order to

[handwritten margin note: Irony of increased income arguments...]

satisfy either the Pareto criterion or, more modestly, the desire to sustain a minimum level of well-being for all generations across time. The task of designing and implementing such an intergenerational transfer mechanism would, in turn, necessitate societal discussion regarding the most desirable method of accomplishing intergenerational transfers, including a discussion regarding the composition of the resources that are to be allocated for the benefit of posterity.

Such a conversation returns us to the challenge of squaring intergenerational ecological needs with liberal political theory. Like Rawls, welfare economists tend to reject resource-specific intergenerational planning in favor of a more diffuse obligation to transfer "an accumulation of capital." One prominent legal economist, for instance, argues that "it is incomplete and potentially misleading to suggest that the present generation does (or does not) have an obligation to a future generation to do one specific thing or another, such as cleaning up the environment, conserving nonrenewable resources, or avoiding accumulation of a large debt."[44] Similarly, recent defenders of discounting assert that "there is no abstract reason to believe that preserving a particular environmental amenity (a forest, a lake) is always better for posterity than other investments that do not involve the environment in particular (expenditures on basic research, reductions in national debt)."[45] Such statements could be multiplied indefinitely.[46] Collectively, they reflect what has been called the weak sustainability conception in the environmental economics literature, a position that harkens back to Rawls's just-savings principle. Specifically, in order to promote nondeclining utility over time, proponents of weak sustainability require only that a portion of the proceeds from resource exhaustion be reinvested in reproducible capital, while remaining essentially agnostic on the market's preference among competing uses of natural resources.[47]

At their core, these arguments depend on a commensurated view of the world, one in which near-perfect substitutability exists among the varieties of human and natural capital. Economists are motivated to accept this perfect substitutability assumption in part because they believe that the attempt to preserve any particular portfolio of natural capital assets will lead to wasteful decision making. But they also are motivated to accept the assumption simply because their separationist enterprise requires it: The social goals of distributive equity and environmental sustainability can only be safely banished from the domain of policy analysis if they can be adequately satisfied through ex-post monetary transfers, as opposed to resource-specific entitlements. At bottom, then, the weak sustainability conception depends on a conviction that any form of capital can compensate for any form of deprivation—that improved productive efficiency, resource substitution, and technological innovation can resolve any particular environmental or natural resource problem that might arise in the future. This is similar to the optimistic disposition described above, in which the fruits of laissez-faire economic policies are believed to inevitably "trickle forward" to future generations. Such optimism is not limited to proponents of the economic approach, but rather reflects a widespread belief in progress within Western intellectual traditions, embraced even by those who are otherwise critical of much of Enlightenment liberalism.[48]

Despite, or perhaps because of, this widely shared outlook, the assumptions underlying the weak sustainability conception tend to receive inadequate scientific attention. This is particularly the case with respect to the assumption that intergenerational equity can be addressed through the transfer of an unspecified resource base. Many physical scientists who have addressed the environmental sustainability question are far less sanguine than economists, political theorists, and other social scientists. They favor instead a regime of strong sustainability, in which government policies limit natural resource uses and other human activities to ecologically determined conditions of sustainability. Under a strong sustainability regime, for instance, renewable resources would be exploited only at a rate that can be repeated indefinitely, such as the natural recharge rate in the case of an underground aquifer. Nonrenewable resources, in turn, would be depleted only at a rate equal to the rate of development of actual substitute resources.[49] Together, these rules of thumb would help to effect a practical realization of Thomas Jefferson's injunction that "[t]he earth belongs in usufruct to the living."[50] The fact that economists and other social scientists instead continue to adhere to the perfect substitutability assumption, even with respect to the fundamental environmental resource of climate stability, is probably not attributable to a disregard by them for the knowledge and credibility of natural scientists.[51] Instead, it likely originates from a fear that accepting the natural scientists' position would undermine the liberal project of avoiding government arbitration among competing conceptions of the good, and the related economic project of avoiding interpersonal welfare comparisons. Reluctant to "break out and model explicitly the consumption of ecosystem services yielded by our stock of natural capital,"[52] analysts instead model the world of their political desires.

Finally, like the rate-of-time-preference defense, the opportunity-costs defense of discounting also suffers from a logical error, even assuming that its other failings could be overcome. Specifically, proponents of discounting hinge the decision whether to conserve natural resources for future generations on the size of the opportunity cost entailed by conservation, when in fact much of environmental policy making is better conceived as being determinative of, rather than determined by, the market rates of return that embody such opportunity costs. If the savings rate for fossil fuels, arable soil, freshwater, wetlands, and other forms of natural capital in part determines the rate of return for all capital—if, in other words, the decision whether to conserve natural resources influences the size of the opportunity cost that supposedly determines whether or not it is optimal to conserve—then the justification for discounting by market rates of return is circular. After all, a major reason that many economists view greenhouse gas mitigation as a "low-yielding" investment"[53] is that they compare it through discounting to a rate of return for capital that takes as given the fossil fuel economy with its laissez-faire climate change policy. The status quo of continued greenhouse gas emissions thus appears to be a better investment than mitigation, in part because the benefits of abatement have been discounted at a rate that assumes mitigation will not occur. Such circularity may be tolerable for decisions of modest practical impact—in which the ultimate outcome might not be affected by the specification of a different reference case of resource rights—but it hardly seems appropriate for addressing the type of large-scale, long-term issues, such as climate change, that increasingly occupy the center stage of environmental law and policy.[54]

Like it or not, then, the need to directly address intergenerational resource equity cannot be avoided: Even within thought experiments involving perfect futures markets and zero impediments to intergenerational bargaining, one still faces the problem of establishing an initial endowment of resources between generations. Every distribution

of resources between generations gives rise to a different market equilibrium, including within that equilibrium a resultant market rate of
interest that will be taken to reflect the opportunity cost of capital within
discounted social welfare policy analysis. We must therefore address the
baseline distribution of resources across generations *first*, before intergenerational efficiency analysis can even begin to get off the ground.
Our failure to perceive this need—our belief instead that the current
market equilibrium somehow reflects a normatively privileged moment
even with respect to the unborn—is indefensible. It is a forward-looking
parallel to libertarianism's attempt to wash clean the present market
equilibrium from its past history of injustice and arbitrariness. Put
bluntly, there is no objective or free-market price that can be identified
for nonrenewable and exhaustible resources such as oil, coal, minerals,
timber, and groundwater—resources that nevertheless lie at the foundation of a substantial portion of all economic activity in industrialized
countries. In the environmental economics literature, the standard wisdom regarding such resources is that the net returns from their extrac-
tion should be reinvested in reproducible capital in order to ensure
sustainability and intergenerational equity.[55] But analysts must have
some prior notion of the "shadow price" of exhaustible resources in
order to determine the amount of reinvestment required. That notion,
in turn, requires making judgments about the policy features of an idealized economy in which "various components of . . . ecological capital"
are maintained above "critical threshold levels which are stipulated to be
preserved to ensure system resilience"[56]—precisely the kinds of judgments that economists instead want to subject to welfare-maximization
procedures using discount rates and other data derived from current,
unsustainable market conditions.

As an illustration of these points, consider the following, seemingly
straightforward question: At what rate should the benefits of future fish
consumption be discounted when using cost-benefit analysis to determine the appropriate level of harvesting from a particular fishery, such
as the Northwest Atlantic cod stocks that once thrived off the coast of
New England? If one's response is the prevailing market rate of interest
for goods, or the rate of return available for alternative public investments, or even some socially determined rate of discount set according
to abstract principles of intergenerational equity, then one is in agreement

with Robert Solow, who argued that "[i]f you don't eat one species of fish, you can eat another species of fish," and that "[t]here is no specific object that the goal of sustainability, the obligation of sustainability, requires us to leave untouched."[57] However its exact rate is determined, the use of a discount factor to fix intergenerational resource rights implies that cod survival is entirely contingent on whether the rate happens to permit survival.

If, on the other hand, one's response to the question is some variation of "at whatever rate results in a sustainably managed fishery," then one is in agreement with the view that distributive judgments regarding certain scarce ecological goods *precede* the choice of discount rate. One accepts, in other words, that the very values to be discounted within the welfare-maximization exercise depend on specification of a background distribution of rights and responsibilities across generations, and that much of environmental law and policy making is concerned precisely with that analytically prior question of resource distribution among generations. To be sure, the fishery example uses a context in which the distributive judgment is relatively tractable and even appears to admit of an "optimal" outcome: Renewable resources should be harvested at a level that ensures maximum sustainable yield. Nevertheless, the same conceptual framework underlies all issues of intergenerational environmental import, including most pressingly the distribution across time of the atmosphere's limited capacity to tolerate greenhouse gas emissions. Discounting cannot resolve these issues. Instead, the question of how natural resources and ecosystem services capacities should be distributed across time remains fundamentally an ethical question, one that will imply a discount rate but that cannot be determined by it.

*

Intergenerational discounting becomes even more problematic when the anticipated impacts of policy proposals on future generations include enhanced human mortality and morbidity. Defenders of the practice tend to assume that "life, risks to life, and enjoyment of life can be measured in money,"[58] an assumption that purportedly enables them to demonstrate the desirability of intergenerational discounting separate and apart from problems of valuing human life. Somewhat defensively, they argue that their procedure does not require the discounting of future lives at all, but rather only the discounting of "costs and benefits,"[59]

[margin note: analytically prior questions of resource distribution]

"people's willingness to pay to avoid . . . risks"[60] or "a monetary amount equal to the willingness to pay to reduce risks to life."[61] They insist that "what is being discounted is the monetary value of the risk itself,"[62] and that the relevant issue is a "technical economics question," rather than "an ethical question about whether discounting is appropriate."[63]

These analysts protest too much. The Pareto criterion is the gold standard for welfare-economic-policy analysis because it is thought to avoid the problem of interpersonal welfare comparisons. That is, each individual *herself* determines whether she is made better off—or at least no worse off—by a proposed policy. Whatever one thinks of such an approach for decision making within generations, it does not translate to decision making between generations, for the simple reason that future generations do not yet exist. Thus, there can be no Pareto criterion in the intergenerational context, at least not when that term is understood to require individuals themselves to determine whether they are made better or worse off by a policy proposal. This is no minor complication: The practice of divining and satisfying the preference functions of future generations must stand on a normative foothold that is separate and distinct from that which typically supports welfarist policy approaches intragenerationally.

In that respect, the most powerful normative argument for discounting—that it will leave future generations with a more valuable stock of resources by paying heed to opportunity costs—is embarrassed by the fact that some members of those future generations are sacrificed in favor of the very alternative investments that are supposed to inure to their benefit. At times, this embarrassment seems to be clear to all but the welfare maximizer himself. One prominent discussion of cost-benefit analysis and climate change, for instance, includes the claim that "whether future generations will accept an increase in the rate of skin cancer or not depends on what they get in exchange for it."[64] Likewise, legal economic analysts confidently assert that "[i]f we could ask future generations whether they would want us to engage in [a particular] project [that fails cost-benefit analysis after discounting], they would prefer that we just invest the money at the market rate of return because they would be better off with such an investment."[65]

Notice what has happened through these claims: The authors have subtly shifted from an individualist to a collectivist conception of the relevant interest holder, so that the same entity ("future generations") suffering the costs of increased mortality also appears to be the one receiving compensation for the loss. This is similar to the conceptual slide, discussed above, that occurs when welfare maximizers discount continuously for time preference, even though human beings only experience temporal impatience beginning with and for the duration of their individual lives. In both instances, future generations are given apparent normative significance as collective entities, even though welfare maximization itself proceeds on the basis of *individualistic* assessments of welfare. This unacknowledged shift from the individualist to the collectivist perspective can lead to absurd results, including the possibility that future generations can appear to be "better off" even when they have been rendered extinct. That is, because lives are monetized and traded along with all other resources, nothing within the standard cost-benefit-analysis framework excludes the possibility that it could appear to be welfare maximizing to kill off every single member of the human race, while the stock of capital that supposedly makes future generations "better off" continues to grow exponentially in bank accounts that will never be withdrawn. The fact that our would-be successors will never collect these funds suggests that our theory has exceeded the bounds of its competency; we have forgotten what John Ruskin called the "one great fact," that "[t]here is no wealth but life."[66]

This point cannot be emphasized enough: All the discounting proponent really demonstrates by appealing to opportunity costs in the life-saving context is the fact that a life lost in the future may be *compensated for* at lower cost than a life lost today. The decision actually to sacrifice the life—and thereby to bring about a situation in which compensation becomes relevant—remains an entirely separate and philosophically more-problematic matter. This separation is made clear in a famous example from Derek Parfit, in which he argues that the harm associated with deformities or other serious physical consequences that we might impose on future generations does not vary with time in the manner of monetary investments.[67] Proponents of the economic paradigm respond to this seemingly irrefutable point by arguing that discounting future deformities to a present value is still appropriate for public policy analysis because "the cost of care or cost of a cure for deformities is likely to go down over time."[68] Even on its own terms, this argument is weakened when evaluated from the standpoint of actual well-being, as opposed to the monetary equivalents of well-being: The same economic forces that purportedly enable a present generation to set aside a lesser amount today to compensate for a harm tomorrow also imply that the amount required to compensate for a harm tomorrow will be greater, due to the declining marginal utility provided by money and tangible forms of compensation.[69] The monetary equivalent of health only remains proportionate with rising income if health can be monetized in the same fashion as any other good, and if the resulting "monetary equivalent" behaves similarly to those goods as income shifts and time progresses. Whatever the validity of this depiction with respect to minor health impacts, it is implausible for death and other extreme inflictions of harm, where the moral and psychological weight of the injury caused to future individuals will bear little relationship to the long-run economic growth rate.

More importantly, the welfare economist's response to the deformity case misstates the nature of the ethical question on the table, which is not how to care for or cure a deformity that has already been suffered, but instead whether to inflict the harm in the first place. By presuming that there is no ethically relevant distinction between living as a nondeformed individual and living as a deformed individual who has received compensation, the economist's response violates the most fundamental ethical precept, that individuals should not be used without their consent as means rather than as ends. Harm is inflicted—future individuals are used and indeed sacrificed without their consent—in order to promote well-being that they will not enjoy. This is why in the intergenerational context one cannot separate the issue of discounting from the issue of valuation. This is also why, when analysts assert that "once the relevant amounts are generated . . . they will be monetary, and they must be discounted,"[70] they succeed only in pushing the important ethical decisions back one step in the analysis to the question of whether and how to monetize life in the intergenerational context.

Consider another thought experiment to help elucidate this point: Suppose that society set up a trust fund from which equal payments would be made to surviving family members or other representatives on account of the death of any individual member of the society. In light of the differing life expectancies that society members would have, the managers of the trust fund quite properly could earmark a lesser amount of money for younger members than for older members. Likewise, they could set aside an even lesser amount for the not-yet-born. In both cases, the later expected death of the individual would allow the lesser amount set aside to grow through alternative investment until payment eventually became due. Notice, though, that the more or less arbitrary fact of being born later in time would not support a judgment that the future life actually is "worth" less than the current life, at least not in any sense that would justify the knowing sacrifice of one life over the other. To see why, suppose that some unforeseen societal emergency actually did necessitate the knowing sacrifice of a given number of lives. Would anyone argue that the youngest members of society should be chosen simply because the trust fund accounting system appeared to show that their lives have the lowest current monetary equivalent value? Suppose further that the society could respond to the emergency by sacrificing a given number of the not-yet-born, perhaps by storing hazardous waste in a container that will leak in a hundred years time and cause a predictable number of deaths. Would the time value of money by itself justify taking those future lives over the currently living? Suppose finally that the society could respond to the emergency either by sacrificing a given number of the not-yet-born or by lowering its current standard of living by 1–2 percentage points of GDP, as many economic analysts estimate we could do in order to reduce significantly the possibility of catastrophic climate change. Now, would it not be clear that the trust fund accounting was entirely beside the point?

It is one thing (a shoddy thing, as chapter 4 argued) to presume individualized consent by members of the current generation to all manner of environmental, health, and safety risks based on controversial revealed-preference studies. It is quite another to presume such consent among future individuals who have neither vote nor voice nor volition to leave our political community and its sphere of impact.

*

Nothing in the foregoing discussion is intended to suggest that analysts are not right to be focusing on opportunity costs when considering intergenerational policy dilemmas, or that a society should flatly ignore the benefits forgone by cautionary environmental policies. Such opportunity costs, however, should not be compounded mechanically into a welfare-maximization exercise, at least not without first asking certain foundational questions regarding intergenerational environmental equity. In what remains the most thoughtful discussion of discounting in the environmental law literature, Richard Revesz similarly concluded that intergenerational decision making should "tak[e] account of" opportunity costs, but not be dictated by a particular discount rate.[71] Economists sometimes contend that no such distinction exists between "taking account of" opportunity costs and accommodating them—mathematically—within the welfare calculus, much as they occasionally deny that a meaningful distinction exists between risk and uncertainty.[72] Such contentions miss their mark, however, for they presume a kind of calculative compulsion that Revesz specifically rejected. Not all rationality is formalized, and despite the frequent claim that welfare maximization is desirable because it encourages comprehensive assessment of outcomes, it is in fact *only* formalized systems, such as cost-benefit analysis, that must, by their very nature, be incomplete. Revesz seemed to recognize this unavoidable analytical dilemma and chose to sacrifice formalized, axiomatic consistency rather than substantive completeness; thus, much like proponents of the precautionary principle and sustainable development, Revesz viewed opportunity costs as but one factor in a pluralistic assessment of intergenerational obligation.[73]

In contrast, proponents of welfare maximization choose to emphasize logical consistency—an intellectual preference that, although they rarely admit it, requires them to sacrifice completeness. As revealed throughout this chapter, one significant way in which welfare-economic-policy analysis sacrifices completeness is through the use of a discount rate that assumes questions of intergenerational environmental equity either may be adequately addressed through ex-post monetary transfers or, more brashly, may be ignored altogether. The defense of these assumptions is unpersuasive: Through discounting, fundamental issues of [intergenerational equity] (which risks and resources, as an ethical matter, should be imposed or bestowed on future generations?) are conflated with the issue of [intergenerational efficiency] (which generation, as a technical matter based on a given rate of discount and presumed distribution of entitlements, does or will derive more utility from the use of a resource?). Through this conflation, future generations are forced to outbid present owners for natural resources by an amount reflecting not only the strength of their needs but also the alternative uses to which all resources, including the "monetary equivalents" of their own lives, could be put during the intervening time periods. This is conceptual confusion. Even if transfers of monetary resources to future generations *are* offered as compensation for this bidding disparity, the discounting procedure still suffers from a basic flaw: The efficiency exercise that determines the amount of compensation due will have proceeded on the basis of a discount rate that assumes away the hard work of evaluating intergenerational equity.

All of these points once were well appreciated even by welfare economists. Founders of the discipline at one time expressed the view that any form of discounting is "ethically indefensible," that it "implies only that our telescopic faculty is defective," and that, at best, the practice serves as "a polite expression for rapacity and the conquest of reason by passion."[74] Even contemporary defenders of welfare maximization at times recognize the need to think more directly about intergenerational environmental justice. Recent commentators, for instance, state that "[i]f respecting future generations means anything, it should mean respecting our best guess as to their wishes and helping them as much as feasible."[75] Nowhere in their analysis, however, do the commentators actually engage the question of what "our best guess as to their wishes" is or how we might go about constructing and operationalizing a process to divine and satisfy those wishes. Instead, they simply assume that our present preferences can be projected onto future generations and that their interests can be discounted as if they belonged entirely to us. This is an odd approach, not only because of its temporal feudalism, but also because the one thing we do know with confidence about climate change is that the future environmental context will be vastly different from the present. The preferences of future generations undoubtedly will reflect these dramatic changes, making them quite different from our own. Thus, in order to truly respect the interests of future generations, we must undertake an engaged effort to anticipate and consider the details of their plight and to provide the specific institutions and resources they will need in order to endure it.

This attempt to specifically anticipate and provide for our unknowable successors should not be taken as a grand threat to our liberal convictions, but rather as a simple acknowledgement that, especially in the intergenerational context, we cannot *not* have a profound influence on the possibilities and predilections of others. More so than perhaps any other crack in the façade of liberal neutrality, the temporal dislocation of future generations challenges the notion that individual autonomy can exist as an unproblematic concept, for it is inevitable that the actions and decisions of presently living individuals, both public and private, will influence the needs and desires of future individuals. As Talbot Page observed, this problem of endogenous preferences is especially acute in the intergenerational policy-making context: "How well the resource base is kept intact—how diminished it will be in biological diversity, how depleted in its soils, forests, groundwater, how crowded in population—will shape our grandchildren's prospects and values and in doing so will shape intergenerational society."[76] In this respect, it is striking that Rawls saw fit to apply the difference principle to presently living individuals—who typically *can* plead or protest to some extent, even within societies that are not characterized by just institutions—while denying such distributive protection to individuals yet unborn, who are most at risk of exploitation and, indeed, who are waiting inexorably outside the presumed conditions of justice.

As chapter 9 argues, constitutional guarantee of some modicum of environmental support for future generations—such as a right to adequate freshwater for sustenance and sanitation or a basic level of climate stability—not only would give the just-savings principle a sound basis in ecology, it also would serve to render future generations more visible to us, closer, and less alien. It would help to ensure that they appear within our social partnership as recognized—albeit unknowable—equals, rather than as mere utility vessels whose survival depends on the results of our welfare calculations just happening to work so that their survival is desirable. We do well by future generations not when we seek in vain to avoid any coercive influence on their life courses at all, nor when we commit to maximize some disembodied stock of welfare, of which they may or may not live to partake, but rather when we provide for their basic needs, when we ensure for them at least "the wherewithal to breathe and to live."[77] The ideal of being open to the unpredictable arrival of an unknown and uninvited guest—the ideal of hospitality which some have taken to be instructive for political theory[78]—does not counsel indifference or neutrality or even tolerance toward the guest. It counsels instead a form of love, which in this context admits of at least some simple commands, for we safely can assume that, whatever else may come, our guest will require food, water, and shelter.[79]

Even if we somehow turn out to be wrong in this prediction—even if the technologies of future generations so remove them from our ecological and biological heritage as to render our provisions useless—we will not have squandered an opportunity to better manipulate our material world. We will have instead gained a continuation of the sense of intergenerational recognition and responsibility that must precede all such material manipulation.

CHAPTER SEVEN

Other Forms of Life

THE TEMPORAL REMOTENESS OF future generations exposes significant blind spots within individualism, both as an analytical method and as a normative commitment to read political questions through the eyes of the individual. These blind spots are perhaps most powerfully demonstrated by what Derek Parfit terms the "non-identity problem": the fact that whatever policy is selected for a given issue may affect the very identity of future individuals.[1] The non-identity problem is related to, but distinct from, the problem of endogenous preferences, which, as discussed in chapter 4, stems from the possibility that environmental policy may have profound effects on the preferences of individuals, including even those previously identified preferences that may have been used to determine the content of environmental policy. The non-identity problem suggests not merely that future individuals' preferences shift as a result of policy choices, but that their very existence is contingent on those choices. Different sets of individuals, in other words, come into existence depending on how we decide. Under such circumstances, we cannot say that particular individuals in the future will be made better or worse off by a policy, only that they will be made.

To the student of complexity theory, this problem of contingent identity is quickly recognized as a manifestation of the profound endogeneity and interconnectedness of complex adaptive systems; to the welfare maximizer, on the other hand, the problem is fundamentally disruptive. Once we acknowledge the fact of contingent identity, our welfare calculus must admit of multiple equilibria, representing entirely different communities of interest; we must therefore develop some normative criteria outside of welfarism to select among them. Indeed, as Parfit notes, the non-identity problem poses deep conceptual challenges to any normative theory that is framed in terms of the rights, preferences, or interests of particular individuals.[2] Such "person-affecting" theories, as Parfit calls them, provide little analytical traction in decision-making contexts where the relevant consequences will be felt by entirely different persons, depending on how the decision is resolved.

At present, our moral and political thinking regarding such "different people choices" is highly immature. It has led, for instance, to the conclusions that we have no obligation to future generations whatsoever,[3] that we have only an obligation to ensure that our choices leave future generations with lives that are minimally worth living,[4] and, more generally, that we "who are currently existing constitute a kind of ontological elite whose interests are to take strict precedence over the interests of everyone else."[5] In order to avoid such unattractive conclusions, Parfit suggests that we need to move beyond simple "appeal to what is good or bad for those people whom our acts affect," and instead develop methods of evaluating "different sets of possible lives."[6] Conceiving of future generations as coherent collective entities, rather than merely as individual lives-in-waiting, is one promising step in that direction. It also bears repeating that the precautionary principle, despite all of its alleged failings, promotes just such a collective conception by establishing a standard of agent-relative environmental responsibility in which human societies and generations can be seen as distinct agents that stand in relations of responsibility and indebtedness to each other. Through such a construction, we can begin to harmonize our liberal individualistic ideals with the reality that some measure of influence and paternalism is simply inescapable vis-à-vis members of future generations.

Nevertheless, prominent defenders of the welfarist approach to law and policy strongly reject this organicist and future-oriented political conception, contending instead that social choice must reflect the interests solely of present individuals "who have the vote and thus can affect the electoral success of current government officials."[7] They reject the

idea of self-conscious collective engagement with the needs and interests of members of other political communities, including even future
members of the instant community. In their view, policy makers should
promote the interests of future generations only to the extent that members of the current generation desire altruistic policies based on their
preexisting preferences.[8] Some theorists go further by denying even the
legitimacy of altruistic preferences, arguing that cost-benefit analysis
should reflect only self-interested preferences in order to truly embody
the welfarist ideal.[9] Views of this bent have strong policy implications, as
made evident in legal economic scholar Eric Posner's response to the
claim that environmental law should grapple with the ethical claims of
other interest holders: "[Government officials] are delegated the power
to maximize the welfare of the current generation of Americans, not the
welfare of all future generations of Americans, and not the welfare of the
current and future population of the world. . . . Just as agencies have no
authority to enact regulations that enhance the well-being of foreigners,
they have no authority to enact regulations that enhance the well-being
of future Americans independently of the preferences of current Americans. *The future is another country.*"[10]

As with welfarism's implicit assumption that the present generation
owns the entirety of the natural resource base, all of the important analytical work for Posner's conception is done in his foundational assumption that "agencies should ignore future utilities just as they currently
ignore the utilities of aliens."[11] If agencies were to follow this approach,
however, they would be ignoring the unequivocal command of numerous statutes that, as noted in chapters 5 and 6, expressly direct agencies
in various contexts to consider the nation's obligations to members of
other political communities and to future generations. Agencies under
those statutes are directed to actively consider and construct the nation's
ethical relation to other communities, not to implement a preexisting
welfare function that may or may not include other-regarding preferences. To put it bluntly, then, the insular and presentist political community conjured by Posner exists only in the theoretical imagination of
certain proponents of cost-benefit analysis: Rather than the future, it is
the welfare economist himself who constitutes "another country."

To be fair, the impulse to elide ethical and political questions surrounding membership is not only felt in welfare economics: Contempo-

[handwritten margin note: But some statutes mandate looking towards future generations...]

rary political philosophy in general tends to focus on the question of what equality demands in terms of its substantive content (for example, political participation rights, primary goods, functionings, capabilities, resources, welfare) and its pattern of realization (for example, equality, sufficiency, minimax, Pareto optimality, Kaldor-Hicks optimality). Much less attention is devoted to the question of membership—the question of who is, or is not, granted standing in the political community that will benefit from the chosen metric and the selected distributive pattern of equality. This oversight is striking given that, as Seyla Benhabib observes, "[t]he question—'who is the we?'—is a fundamental political question."[12] Indeed, it might be considered *the* fundamental political question, as its answer at least implicitly works to structure and constrain all other political judgments.[13] Even in the radical view suggested by Posner—in which the state seals its borders and ignores its future, seeking only to implement the preferences of existing citizens—the question of membership still haunts the community in the form of well-known paradoxes of social choice that have been identified by Amartya Sen and Kenneth Arrow.[14] Although demonstrated through intricate formal proofs, Arrow and Sen's results also can be seen to represent echoes of the foundational paradox of constitutional democracy; that is, the originary moment of co-legitimation, in which a people is said to have expressed its democratic will to form a political community through a constitutional irruption that simultaneously is said to have received the people's assent as a mechanism of will-recognition. Legitimation of the constitution demands popular sovereignty, yet the popular sovereign's will must be discerned through principles of constitutional law that already somehow have been legitimated. This paradox cannot be resolved analytically; instead, it must simply live on in the continuing and never completely successful efforts of the community to overcome its foundational tension, to demonstrate the continued vitality of the performative utterance, "We the People," which first constituted the political community as a community even as the utterance represented the community as already formed.[15] In this way—by recognizing, rather than denying or ignoring, the fact that membership always is at issue in policy making and that every instance of social choice raises anew the question "Are *we* 'the People'?"—democratic constitutionalism persists despite its paradoxical origin.[16]

Although already explored in the previous two chapters, the question of membership takes on a particularly sharp rendering in the context of other forms of life. As noted above, cost-benefit analysis is typically premised on a liberal conception in which only humans—and, more specifically, only presently living individual humans—are capable of holding interests. For that reason, the value of nonhuman life-forms is acknowledged only to the extent that identifiable human individuals value those life-forms, generally through revealed preference or contingent valuation studies. As described in chapter 4, this value must take a finite, monetized form, so that the resulting "worth" of any aspect of the natural world is readily exchangeable for any other item of worth. Even from a purely welfarist vantage point, in which the value of nonhuman life can only consist in its instrumental use or sentimental worth to humans, one might question why welfare economists feel themselves empowered to evaluate the merits of this homogenizing assumption. Why not, instead, defer to natural scientists who study directly the role of natural phenomena in lived experience and who have adopted increasingly desperate attempts to express their collective judgment that "[h]uman beings and the natural world are on a collision course"[17]? As described in chapter 6, the distinction between "weak" and "strong" conceptions of sustainability in environmental economics hinges essentially on whether the policy analyst is willing to defer to the claims of natural scientists that some natural resources should be treated as beyond measurement and trading, as lexically prior to the framework of market exchange.

In fact, the *only* value that is given lexical primacy in the standard approach—and that is therefore exempted from the demand that its worth constantly be demonstrated in order to avoid being displaced—is the economists' preference for preferentialist assessments of value. What this means in practice is that the value of any ecological good or service is only that amount of monetary commitment to the good or service that can be identified in the behavior of individuals by professional economists using the established tools of the guild. This approach begs deeper justification: Neither the individuals who are subjected to preference-elicitation studies nor the economists who interpret the results are necessarily well versed in the underlying scientific questions regarding the role, resilience, and replaceability of natural resources and ecosystem services. For that reason, the most informed preference of individuals—their "true preference," if such a term is insisted upon—may be precisely *not* to have collective decisions regarding the ecological fate of the planet made to depend upon their ill-informed and unstable consumerist preferences.[18]

More fundamentally, the valuation practices of welfare economics appear to succeed as an analytical matter only because they have assumed that the ethical and political boundaries between life-forms—between humans as interest holders and nonhumans as objects of valuation—are preestablished and noncontroversial. One way of understanding the project of environmental law, however, is precisely as an effort to subject those boundaries to renewed examination and critique. The NEPA, for instance, established as a foundational goal the pursuit of "conditions under which man and nature can exist in productive harmony. . . ."[19] The ESA, as interpreted by the Supreme Court in *Tennessee Valley Authority v. Hill*, likewise reflected "a conscious decision by Congress to give endangered species priority over the 'primary missions' of federal agencies."[20] In both instances, the desire to position nature in a relationship of mutual "harmony" and "to halt and reverse the trend toward species extinction, whatever the cost"[21] suggests at least the beginnings of a process of endowing nature and nonhuman life-forms with a legal status above and beyond that of mere object. Formal designation of nature's incommensurability was intended not merely to serve the anthropocentric purpose of ensuring conservation while humanity developed better knowledge of nature's instrumental worth, although that surely was one argument deployed in favor of such laws. Instead, the prioritization of nonhuman life-forms and environmental interests also was intended to commence a process of deliberate reconsideration of humanity's relation to the rest of the natural world and to imbue that relation with a renewed sense of responsibility.

Such a process of collective self-examination simply cannot be countenanced within the welfare-economic framework, since it takes as given the relevant sphere of interest holders and, indeed, specifically denies those interest holders a vantage point from which to reassess the foundational membership decisions that, although resolved offstage, nevertheless continually bind them together into a distinctive cost-benefit community. In that respect, Charles Taylor's observations regarding the seductive effect of what he called the "modern social imaginary" apply with especial force to the cost-benefit community, since welfare economics demands the expression of all forms of value within its monistic language. As Taylor wrote, "once we are well installed in the modern social imaginary, it seems the only possible one, the only one that makes sense. After all, are we not all individuals? Do we not associate in society for our mutual benefit? How else to measure social life?"[22] Equally alluring is the welfare-economic framework: After all, are tradeoffs not inevitable? Do we not approach every decision with an understanding, rough perhaps, of how it will affect the well-being of people? How else to measure social life?

As the remainder of this chapter argues, the welfare-economic insistence on regulating from nowhere—perhaps the ultimate expression of the modern social imaginary—is now dangerously ignorant of the questions of membership that loom offstage. Not only does the welfare-economic vantage point unduly truncate our relation to the natural world by presuming only a mode of ownership and control, rather than of intersubjective recognition and productive harmony, but it also imposes a myopic lens that will fail to recognize, let alone resolve, the ethical challenges raised by genetic engineering, cloning, and other potentially radical interventions into the formation of life. Ultimately, those technologies will offer the prospect of redefining even the "presently living individual," which welfare economics axiomatically takes to be the only locus of value in the cost-benefit community. In that sense, the question of whether to permit the engineering of superhuman—or, more worryingly still, subhuman—classes of individuals will not be resolved with reference to whether it will be good or bad for them, or even for us. The question rather will be "Are *they* 'We'?"

*

Readers well versed in the intellectual history of the animal welfare movement and the related, although occasionally opposed, movement on behalf of the "rights of nature"[23] will recognize something peculiar about the preceding argument: Despite the alleged willful blindness of welfare economics to the moral and legal status of nonhuman life-forms, the most influential arguments on behalf of animal welfare legislation over the past two centuries have emerged from utilitarian thinkers—the close philosophical cousins of welfare economists. While the latter group typically employs preferentialist accounts of value in which only humans are deemed capable of revealing or expressing their desire for particular outcomes, the former group often engages more directly with the psychological or hedonic basis of well-being. Thus, in contrast to the narrow cabining of value found in welfare economics, the utilitarian philosophical approach invites consideration of whether nonhuman life-forms experience mental states that might be "counted" for purposes of optimization. The classic statement along these lines belongs to Jeremy Bentham, who saw in utilitarianism not merely a method of moral calculation, but an indispensable tool for overcoming prejudiced habits of thought and social practice. His statement with regard to nonhuman life-forms is worth quoting at length:

> The day *may* come, when the rest of the animal creation may acquire those rights which never could have been withholden from them but by the hand of tyranny. The French have already discovered that the blackness of the skin is no reason why a human being should be abandoned without redress to the caprice of a tormentor. It may come one day to be recognized, that the number of the legs, the villosity of the skin, or the termination of the *os sacrum*, are reasons equally insufficient for abandoning a sensitive being to the same fate. What else is it that should trace the insuperable line? Is it the faculty of reason, or, perhaps, the faculty of discourse? But a full-grown horse or dog is beyond comparison a more rational, as well as a more conversable animal, than an infant of a day, or a week, or even month, old. But suppose the case were otherwise, what would it avail? the question is not, Can they *reason?* nor, Can they *talk?* but, Can they *suffer?*[24]

Bentham thus offered two notions that would become critical to the animal welfare movement over the ensuing decades: first, the notion that liberalism by its nature promotes an outwardly expanding sphere of interest holders in such a way that individuals regardless of race, gender, class, religion, sexuality, disability status, and so on should be regarded as full and equal members of the political community; and second, the notion that such a progressive expansion of liberalism's community need not be inherently limited to humans, especially when close consideration

expansion of included class...

is given to the various attributes that are conventionally thought to mark humanity off as an exclusive locus of value, yet that, upon inspection, turn out to be shared by at least some nonhuman life-forms. The line instead is superable, at least to the point of sentience.

Although often intermixed with strategically anthropocentric arguments—such as the claims made by Locke and Kant, among others, that cruelty to animals should be prohibited because it conditions the human psyche to more readily commit similar atrocities against humans[25]—Bentham's notions nevertheless continue to underwrite the most influential arguments made in support of moral and legal standing for nonhuman life-forms, including Peter Singer's modern classic *Animal Liberation*. In his early writing on the subject, Singer first complained that "the problem of equality, in moral and political philosophy, is invariably formulated in terms of human equality,"[26] rather than in the more basic terms of whether any particular living entity qualifies as an interest holder. Rather than endorsing the hedonically inclusive approach suggested by Bentham, dominant schools of philosophy in Singer's view simply assume that nonhuman life-forms fail to meet whatever conditions are thought to make possible relations of justice: Whether because of a lack of rough parity in power or resources believed necessary to motivate cooperative political behavior,[27] because of a perceived inability to reason or otherwise engage in the formation of a voluntary social contract,[28] because of a failure to live up to the liberal demand for self-expression of needs and interests,[29] or simply because of an unexamined assumption that "[a]nimals count, but only insofar as they enhance wealth,"[30] nonhuman life-forms have not figured in most attempts to establish the contours of a just political community. Instead, liberal theorists typically follow David Hume in holding that the circumstances of justice attain only between humans (and often only between presently living humans who have eyed each other across nature's battlefield and sensed that dominance through violence cannot reliably be secured).[31]

Hume's contrast with Bentham in this respect could not be sharper. In Hume's words,

> [w]ere there a species of creatures intermingled with men,
> which, though rational, were possessed of such inferior
> strength, both of body and mind, that they were incapable of all

resistance, and could never, upon the highest provocation, make us feel the effects of their resentment, the necessary consequence, I think, is that we should be bound by the laws of humanity to give gentle usage to these creatures, but should not, properly speaking, lie under any restraint of justice with regard to them, nor could they possess any right or property, exclusive of such arbitrary lords. Our intercourse with them could not be called society—which supposes a degree of equality—but absolute command on the one side, and servile obedience on the other. Whatever we covet, they must instantly resign. Our permission is the only tenure by which they hold their possessions; our compassion and kindness the only check by which they curb our lawless will; and as no inconvenience ever results from the exercise of a power so firmly established in nature, the restraints of justice and property, being totally *useless*, would never have place in so unequal a confederacy.[32]

In response to such blunt declarations of the (non)status of "inferior" beings, Singer's strategy followed the twofold approach of Bentham. First, he sought to reawaken individuals' ethical sensitivities by pointing out that numerous classes and groups of human beings have historically been denied recognition as interest holders based similarly on perceptions that their "intrinsic" or "natural" endowments precluded them from full participation as equals in the political community. By positing that a society of laws must "suppose[] a degree of equality" already existent among its members, Hume enabled the wholesale exclusion of those individuals who were believed to be inferior according to dominant naturalistic understandings. Politics, in that sense, was mere ontology: Any life-form deemed lacking in strength or reason according to philosophical presuppositions regarding the nature of being simply fell outside the realm of rights or interests.

Having shown the perniciousness of this construct through its historical use against women, slaves, and other groups, Singer's second move was to follow Bentham in holding that a more basic and appropriate condition for membership in the community of concern is the mere capacity to experience pain and pleasure. Specifically, he argued that "the limit of sentience . . . is the only defensible boundary of concern for the interests of others."[33] For Singer, the value of this approach is somewhat heuristic in that it forces upon humanity an awareness of its own nondistinctiveness: "It is only when we think of humans as no more than a small sub-group of all the beings that inhabit our planet that we may realize that in elevating our own species we are at the same time lowering the relative status of all other species."[34] More ambitiously, the utilitarian framework also aims to impose an analytical rigor that will resist ontological prejudice of the sort exemplified by Hume. Thus, to emphasize the actual hedonic impact of "lowering the relative status of all other species," Singer gathered a substantial and disturbing battery of empirical evidence regarding both the ability of nonhuman animals to suffer and the degree to which they do, in fact, suffer in contemporary agricultural systems, research laboratories, and other settings. Furthermore, in the most controversial aspect of his work, Singer argued that with respect to many characteristics, chimpanzees and other high-functioning animals actually outperform certain members of the human species, such as persons with severe mental disabilities, anencephalic infants, or elderly individuals suffering dementia. To endow the latter with moral standing and legal rights but not the former therefore seemed inconsistent and unjustified, at least from Singer's unblinking utilitarian perspective. Either the case for animal protection should be seen as stronger than it conventionally is or, he argued, the case against infanticide and euthanasia should be seen as weaker.

*

Despite the great intellectual interest and political activity generated by Singer's arguments, the problem remains one of utilitarianism's inherent conceptual limitations. As discussed in chapter 2, when everything is made to depend "on what the facts turn out to be"—as Cass Sunstein and Adrian Vermeule put it in the context of their argument in favor of capital punishment—then the essential elements of agency and responsibility appear to recede from view. Morality becomes a calculus without a calculator, since the utilitarian framework denies the significance, or even the relevance, of any particular entity's point of view, including the very entity engaged in decision making. No doubt in some contexts, such as when dominant agents are ignoring widespread suffering, the deployment of utilitarianism appears to have a distinctively progressive tilt. For instance, by linking the argument on behalf of animal protection solely to empirical facts about the world—namely, "Can they suffer?"—the utilitarian strategy seems to reduce the risk that collective decision making will be unduly influenced by human chauvinism. All that seems required in order to make the case for animal welfare protection is a scientific demonstration that nonhuman life-forms experience pain and pleasure consistent with the basic utilitarian value framework. Once established, this empirical showing promises to unfold without further elaboration into an airtight political program for the inclusion of non-human life-forms within the optimization calculus.

Utilitarianism's essential shortcoming, however, lies precisely with this enticing but perpetually unfulfilled promise that judgment can be programmed, that more precise methods of consequential assessment and more reliable techniques of valuation can fully map the terrain of ethical and political determination. The utilitarian vision fails to appreciate that the truths held by a society are actively produced via human languages, customs, procedures, and institutions. They are, accordingly, contingent, not only in the Kuhnian sense, of being always, per the scientific method, subject to redescription, revision, and occasional wholesale abandonment by expert epistemic communities, but also in the more constructivist sense that their generation occurs within social, economic, and cultural contexts that influence the very nature of the questions asked and the conclusions reached by scientific inquiry. As Bruno Latour more succinctly put it, "No science can exit from the network of its practice."[35] Such a post-positivist understanding of science challenges the utilitarian case for animal protection in two related ways: first, by underscoring the obvious but often ignored point that empirical inputs into utilitarian calculation are themselves social products subject to negotiation and contestation; and, second, by upsetting the various foundational categorical distinctions—such as those between nature and culture or between science and politics—upon which the utilitarian framework has been built.

Broadly speaking, the Enlightenment tradition out of which utilitarianism emerged perceives science as holding exclusive authority to speak for and through nature, while politics is depicted as the exclusive medium through which a self-contained and self-sufficient society governs itself. Much like the levees in New Orleans, this conception has eroded under the heavy spilling over of its categories. As examples such as global climate change or animal cloning make unmistakably clear, society depends on nature, nature is constructed by society, politics is informed by science, science is shaped by politics, and so on in endless permutations, so much so that it is no longer appropriate—indeed it never was appropriate—to treat them as separable domains. They are, at every moment, already each other. Accordingly, the double-delegation of the Enlightenment—in which so-called sociopolitical problems are delegated to professional politicians who are thought to faithfully represent democratic will, and so-called biophysical problems are delegated to professional scientists who are thought to faithfully reveal nature's essence—is unstable, as is the welfare economist's even more ambitious attempt to delegate all problems to a single unified framework of risk-assessment-and-cost-benefit analysis.

It is precisely in this decentered context that utilitarians such as Singer hope to use an empirical demonstration of suffering by animals to make the case for their moral standing and eligibility for legal protection. Future genetic engineering techniques will reveal the shortcomings of such an approach, as humanity will be asked to consider the ethicality not simply of how it treats a life-in-being or a life whose biological constitution is thought to be only marginally under humanity's control, but of how it goes about creating entirely new forms of life. With what characteristics and capacities will those new lives be endowed? Faced with that question, we will no longer be able to rely on Singer's philosophical strategy of comparing human and nonhuman life forms according to some naturalized distribution of attributes in order to determine whether an entity qualifies as an interest holder. Instead, we will be equipped to design new life-forms that either meet or fail the tests for recognition that we have established, whether those tests involve the capacity to feel pain, to perceive an external environment, to communicate with other beings, or to exhibit any other empirically demonstrable characteristic. Through technical means, we will be able to bring those life-forms within or without the sphere of interest holders; the extent of their suffering will therefore depend not on supposed ontological givens, but on moral will stripped of theoretical apology.

Any apparent difference between engineering new life-forms and treating preexisting life-forms can be seen as yet another manifestation of the act-omission distinction, one that may be just as misleading here as elsewhere. In the moment of engineering life, we seem to suddenly appreciate anew the ethicality of our relation to other life and its resistance to formulaic resolution. Thus, the utilitarian strategy of examining characteristics for empirical "fit" with the interest-holder archetype suddenly sputters: Neither science nor ontology can ground our ethics, because we ourselves are designing the package of attributes that characterizes the life-form and that determines its eligibility for inclusion in the optimization calculus. In a sense, though, these powers of determination are always present. Either through the manner in which the test of recognition itself is established or through the inevitable social embeddedness of scientific knowledge or philosophy, the utilitarian strategy for expanding the sphere of interest holders is always dependent on human discretion and agency to a greater extent than its proponents tend to acknowledge. After all, right alongside those followers of Bentham, who seek to widen the circle of concern through moral science, there also have been followers of Herbert Spencer, who seek to keep the circle tightly closed through much the same means.[36]

Again, utilitarianism appears to be Janus-faced, capable both of supporting progressive policies and of undermining all ethical sentiment. Perhaps the key insight to be gleaned from utilitarianism's dual aspect is that the methodology functions best as a method of recognition rather than of optimization: When utilitarianism seems to be supporting normatively desirable policies, it is often because the approach is being used to demand that officials and their publics consider beings who are either ignored or devalued by dominant frameworks of understanding. When utilitarianism itself becomes dominant, however, it is incapable of limiting its language to a proper zone of application. It creeps out into the wider social imaginary and undermines the basis on which any interest holder—or any interest—is deemed distinctive and worthy of especial consideration. Recognition, then, may be the most worthy analytical function of utilitarianism, one that could help us determine when the framework should be enlisted in service of collective self-governance. Of course, as a mechanism of recognition, utilitarianism faces many competitors, including the much-maligned principle of precaution.

Perhaps in light of these difficulties, other philosophical contributors to the animal welfare literature have attempted to make the case for protective policies from a more deontologically oriented perspective.[37]

To these theorists, nonhuman life-forms should be regarded as worthy of protection, not because of some psychological datum, such as their capacity to suffer, but rather because of the more basic idea that nonhuman life-forms hold a unique point of view in the world, one that cannot be entirely subsumed within the point of view of others. Moral significance is thus conferred not through empirical evaluation of the capacities of other life-forms, but rather through an a priori assertion regarding their status as inassimilable holders of particular biographies. In the words of one leading contributor, from this perspective, "moral standing or moral considerability turns upon whether a creature is an experiential subject, with an unfolding series of experiences that, depending upon their quality, can make that creature's life go well or badly."[38]

When put this way, however, it becomes clear that the deontological approach is in danger of devolving into the same kinds of considerations that motivated Singer and his followers. The idea of holding a distinctive point of view in the world, or of having a set of experiences that might constitute a particular biography, seems to depend on at least some assumptions about what it is like to be that other life-form—assumptions that invite empirical analysis and critique. At the very least, the approach seems to require that it be conceptually plausible to regard the other life-form as something from inside of which one could look out, as something the perspective of which one could occupy in order to motivate an ethic of guardianship.[39] Thus, although the deontologically inspired approach to animal protection may appear to be less contingent on psychological conditions and other empirical facts once it is up and running, it still seems to depend on such considerations as part of the initial staging decision regarding which entities are and are not entitled to membership in the "Kingdom of Ends."[40]

This conceptual dilemma is somewhat reminiscent of the one noted in chapter 5 involving the awkward attempt by political theorists to address global justice concerns through tests that evaluate whether states are sufficiently well ordered to achieve recognition as sovereigns and to be treated as ends in themselves rather than as objects of protection or plunder. In both cases, the evaluator tends to ignore his own affirmative role in affecting the outcome of the evaluative process, both in the sense of at least partially prefiguring the destiny of the objects of evaluation through construction of the standard for recognition and in the sense of more directly affecting the characteristics and possibilities of the entities being examined. Just as genetic engineering makes unavoidably plain humanity's role in shaping the life course of other beings, global phenomena such as the adverse effects of climate change underscore the role of developed countries in significantly affecting the conditions under which developing countries must strive to prove themselves capable of governing and providing for their populaces—to become sovereign in the eyes of others. Again, the expansive reach of our agency is beginning to confront us directly, upsetting the categories with which we have previously sought to rationalize the world and to constrain the bounds of our responsibility.

*

Not even the human category will be immune from this process of conceptual rending. At present, much of the ethical and political debate regarding cloning, germ-line genetic engineering, and other advanced biological technologies simply rehearses long-calcified arguments and positions from the abortion debate. With time, however, an even deeper set of philosophical challenges will be seen to arise from the advent of biomedical mastery.[41] In the case of abortion, the question popularly has been taken to be one of whether and when a fetus "counts" as a life-form worthy of protection as a human subject. As Ronald Dworkin has persuasively argued, the question can perhaps better be seen as one of whether human dignity suffers more from the cessation of a potential life-in-being or from the restriction of a woman's biological self-determination.[42] In either case, however, the ontological category of human has not been taken to be seriously at play; most at issue is a question of how early into gestation the human category's penumbra may reach or how seriously the noncorporeal moral dimensions of the human may be undermined by the practice of abortion procedures within a society. By contrast, in the case of coming genetic-engineering technologies, the material constitution of the human itself will become an object of direct and considerable manipulation. As the philosopher David Heyd notes, when faced with such possibilities, we will no longer have recourse to notions of an essential "nature" or "dignity" of human life in our ethics, for the agency we utilize in constructing such notions will become utterly conspicuous.[43]

Many recoil at such a prospect. For instance, President Bush in his 2006 State of the Union address called for a ban on the development of "human-animal hybrids," seeing a danger that such technologies would erode societal belief in the "matchless value of every life."[44] Likewise, the Declaration on the Responsibilities of the Present Generations Towards Future Generations, adopted by the General Conference of the United Nations Educational, Scientific and Cultural Organization in 1997, states both that "[t]he human genome, in full respect of the dignity of the human person and human rights, must be protected," and that "the nature and form of human life must not be undermined in any way whatsoever."[45] Like the speciesism decried by Singer, however, such puritan humanism is difficult to defend analytically, at least not without inviting the same kinds of slippery slopes that were exploited by Singer in his arguments regarding animal welfare, infanticide, and euthanasia. Even the Enlightenment's most insightful contemporary defender, Jürgen Habermas, falls prey to this difficulty in his effort to oppose genetic engineering of the human. In a lecture published as *The Future of Human Nature*, Habermas argues that "ruthless" genetic manipulation would "blur the intuitive distinction between the grown and the made, the subjective and the objective."[46] He fears that such blurring would in turn disrupt the "ethical self-understanding of the [human] species which is crucial for our capacity to see ourselves as the authors of our own life histories, and to recognize one another as autonomous persons."[47] In Habermas's view, the monumental cultural achievement of the Enlightenment, which permits individuals to see themselves as autonomous, self-directed beings rather than as mere ontological givens, would be undone by widespread awareness among humans of having been engineered, of having been quite literally and specifically made to be the product of a set of choices and desires of others.

Habermas foresees the Enlightenment's undoing through this scenario precisely because, in the case of the radically engineered human, "an important boundary has become permeable—the deontological shell which assures the inviolability of the person, the uniqueness of the individual, and the irreplaceability of one's own subjectivity."[48] Having been determined—in the sense of having been produced not just from others' genetic material but from others' precise and detailed decisions, with elements of chance and mystery reduced to a technically attainable minimum—the engineered human becomes "foreign to the reciprocal and symmetrical relations of mutual recognition [that are] proper to a moral and legal community of free and equal persons."[49] Hence, Habermas returns to liberalism's familiar concept of the circumstances of justice, but in his hands the concept is not used as a filtering device to evaluate fitness for membership in the political community. Rather, it offers a normative ideal that must be protected from empirical assault, lest the spirit of individual autonomy be driven from our collective aspiration. Concretely, this means prohibiting deliberate engineering of the human whenever the goal of engineering is not strictly "therapeutic" in nature. To Habermas, therapeutic engineering is permissible only because the recipient of care may regard his construction as having still been directed toward the ideal of autonomy: The clinician's "attitude" will have been one of guardianship or trusteeship of the patient's interests, operating on the basis of ethically infused norms of presumed consent. In so-called enhancement cases, on the other hand, the patient will know that her construction resulted solely from the instrumental engineer's desire to implement his own conception of human optimality.[50]

Despite its admirable nuance and sense of urgency, Habermas's argument suffers from two limitations. First, as in the case of genetic engineering of nonhuman life-forms, his objections understate the degree to which all life-forms—human and nonhuman—have long been subjected to influence and control of a similar, albeit less technically precise nature. In that sense, the "deontological shell" of singularity and self-authorship that Habermas believes to be threatened by genetic engineering is already widely compromised, most obviously by the critical mating decision of one's genetic parents (which can be more self-aggrandizing than Habermas allows), but also by myriad cultural and environmental factors that are imposed upon the individual just as assuredly as genetic design criteria and that frustrate any attempt to fix a conceptual line between genetic "therapy" and "enhancement." Second, to the extent that cognitive awareness of one's heteronomy really is seen to be a threat to the Enlightenment subject and to the political community that it enables, as Habermas fears, then genetic engineering will offer a ready fix: Individuals can simply be engineered to lack awareness regarding their heteronomy, thereby preserving the "ethical self-understanding" that Habermas regards as essential to political morality while still allowing the continuation of technology's "ruthless" advance.

Voilà.

[margin note:] Deontological shells already compromised

[margin note:] Engineer around awareness

Of course, it would constitute a profound insult to Habermas and to the emancipatory legacy of the Enlightenment for consciousness of autonomy simply to be "given" to human subject/products in this manner. The point of the example is not to commend it but rather to emphasize the limits of conventional philosophical approaches to these questions. Whether in the form of Bentham's utilitarianism, Hume's circumstances of justice, or Habermas's deontological shell, each of the examined approaches fails to concede that the other's subjectivity is both fundamentally inaccessible *and* widely exposed to our actions and decisions. To decide to acknowledge another life-form as a source of ethical obligation is not to examine a set of empirical descriptions about that life-form in order to see whether it measures up to some predetermined depiction of the interest holder. It is instead to find ethical obligation precisely in the unknowability of the other's interior world. Likewise, to decide to regulate the engineering of the human is not to preserve an otherwise pure and unproblematic realm of individual autonomy. It is instead to acknowledge that our commitment to human rights at present is far too frail and unrealized for us even to be confident that the human will not be purposefully *diminished* by genetic engineering technologies rather than, as Habermas fears, enhanced. After all, what ethical and political concepts do we have available at present to disfavor the creation of a class of subhuman workers—instrumental beings designed precisely so as *not* to meet whatever cognitive, psychological, or other criteria we have established for equal membership in humanity's community of concern? Following Hume, Hobbes, and other founders, we have designed the circumstances of justice in our political philosophy so that they generally exclude from full consideration members of other countries, other generations, and other forms of life. Why would we not be equally opportunistic when it comes to designing those others directly?

*

The first step toward an honest philosophical encounter with other life-forms is, as David Wood put it, the step back. As he wrote, "[t]he step back marks a certain shape of philosophical practice, one that does not just resign itself to, but affirms the necessity of, ambiguity, incompleteness, repetition, negotiation, and contingency."[51] Embracing indeterminacy in this manner is not to be thought of as an abandonment of analysis but as a resumption of responsibility. As continental philosophers such as Emmanuel Levinas and Jacques Derrida have observed, responsibility depends on "a responsiveness or openness to what does not admit of straightforward decision, to what exceeds conceptualization."[52] Hence, admitting the inevitability of meaning's excess over understanding is the very beginning of responsibility; it is the surrender that enables analysis to yield to decision, decision to action, and action to consequence. Utilitarianism attempts to reverse this sequence by defining human responsibility as that which can be consequentially assessed. The temptation to yield to such seemingly self-executing programs can be great, particularly when a more-honest encounter would bring us face-to-face with suffering that can never be rationalized.

To honestly encounter other life-forms—including the life-forms we choose to call human—we must remain open to the infinite possibilities of their existence. We must reawaken a primordial sense of awe and incomprehension regarding the other's being, one that is infused with an awareness of our unfulfillable responsibility rather than with a desire for programs of thought and action that are total in their reach. Our reawakening must go beyond Singer's historical lessons regarding the precariousness of the project of liberal justice; it must embrace the more revolutionary insight of Levinas's ethics as first philosophy. Unlike the approach of philosophers such as Hume or Bentham, which works to prefigure the destiny of the other by making ontology prior to ethics and politics, Levinas sought to place ethics at the foundation of all systematic thought. Specifically, he sought to recall an awareness of the other's unknowability—"the way in which the other presents himself, *exceeding the idea of the other in me*"[53]—and to take this transcendence as the basis of all subjectivity. On Levinas's account, we do not enter into the world as self-possessed liberal subjects, already equipped with capacities of reason and discourse that enable voluntarily determined justice relations, should we happen to encounter beings worthy of consideration rather than domination. Instead, we emerge into consciousness already hostage to the gaze of the other's face, already placed under an infinite ethical demand.

For Levinas, subjectivity thus begins as awareness of an other who "individuates me in my responsibility for him."[54] The face of the other issues without words a most basic command—"Do not kill me"—and

wakes us to a world in which we inevitably fail our most essential responsibility. Prior to that moment, our consciousness is simply the shared
consciousness of life's infinite value; afterward, as we have seen, our
consciousness tends to deny infinitude and instead to endorse systems
of thoughts and programs of decision that appear to rationalize life and
death. Thus, in Levinas's view, moral and political philosophy must be
reworked from the ground up so that we never deny the infinite value of
life, so that we continually and solemnly confess our failure before the
unsatisfiable demand of the other.[55] Returning to the spectrum of normative ethical theories set out in chapter 1, we might say that Levinas agrees
with the significance of the metaethical task of understanding agency,
separate and apart from the task of selecting a persuasive ethical theory to
guide the conduct of the agent thus understood. In Levinas's rendering,
though, the ethical agent is discovered rather than constructed or analyzed; she opens her eyes to a unique personal responsibility that already
exceeds the content of any theorized ethics or politics.

Ethical agent is discovered, rather than constructed

This individuating encounter with the other should not be thought
to occur in secular time. Well before the modern social imaginary became our synoptic lens for understanding the world, we were familiar
with a notion of sacred or heroic time that does not, in Levinas's words,
"temporalize in a linear way, does not resemble the straightforwardness
of the intentional ray."[56] Rather than a linear unfolding of actions and
consequences, which invites consideration almost exclusively of causal
relations, the time of Levinas's ethics "makes a detour by entering into
the ethical adventure of the relationship to the other person."[57] Because
of this "detour," the call to responsibility issued by the other need not be
seen as emerging from any particular dependent, such as a fellow citizen, a representative of the current generation, or a member of the
human species. Indeed, each of those conceptual distinctions presupposes some ontological categorization of the other and thus fails to remain open to otherness. The ontology that precedes ethics and politics is
violent, and any effort to promote regard for others using those same intellectual means will be self-limiting. Absolute alterity instead demands
absolute openness. The call to responsibility therefore should be heard
to issue from a face outside of secular time. It issues, for instance, from
those faces referred to in the Iroquois Nations Constitution as lying "yet
beneath the surface of the ground—the unborn of the future Nation."[58]

It issues from the unseen victims of natural disasters that occur across the globe, the victims who, in Adam Smith's famous example of an earthquake in China, are conventionally thought to fall outside the realm of ethical concern due solely to their position of geographic remove.[59]

Although Levinas himself wavered on this matter, the call to responsibility should even be heard to issue from the face of animals, for in the ancient time before names and taxonomy, such life-forms were not masked by philosophical presuppositions regarding either their capacities or our responsibilities.[60] Once named, the animal is objectified and made available for possession, exchange, and consumption, in the manner presupposed by welfare economics. Prior to objectification, however, there is simply a gaze of life. This gaze is not mysterious: Anyone who has been caught unawares by an animal—who perhaps found themselves being eyed during a hike by an elk that saw them long before their approach, or who perhaps, like Derrida, simply found themselves after a shower standing naked before their housecat and felt a disorienting sense of being seen by a silent other—would share Derrida's reaction that "[n]othing can ever take away from me the certainty that what we have here is an existence that refuses to be conceptualized."[61] That is, unlike countless philosophers who "have taken no account of the fact that what they call animal could look at them and address them from down there, from a wholly other origin,"[62] the person caught naked before his cat cannot help but recall the ancient time before subjects and objects, before persons and things. He cannot help but recognize, even if only for a moment, that the animal before him—the animal seeing him from its own vantage point—*has* a vantage point, one that cannot be reduced to a name, a description, or a location in a hierarchy of being. Thus, he cannot help but join Derrida in asking, "Who was born first, before the names? Which one saw the other come to this place so long ago? Who will have been the first occupant, and thus the master? Who the subject? Who has remained the despot, for so long now?"[63]

Derrida naked in front of his housecat [margin annotation]

*

As Matthew Calarco observes, the basic flaw with the conventional philosophical approach to the moral status of nonhuman life lies in its framing of the question: It "proceeds as if the question of moral consideration is one that permits of a final answer," rather than proceeding "from the possibility that *anything* might take on a face" in the Levinasian sense, and that "we are further obliged to hold this possibility permanently open."[64] By placing ethics prior to ontology—by placing the onset of self-presence at the moment of recognition of the other's unknowable but unmistakable needs—Levinas denies the subject a position of philosophical complacency. The difference with the Kantian-inspired animal rights approach in this respect is subtle but significant. Subjectivity is not assigned to the other based on an assessment that the other holds a point of view consistent with our criteria for considerability. Rather, subjectivity is assigned *to us* in the form of a piercing recognition of responsibility that precedes self-awareness, a responsibility that can neither be bounded nor satisfied. That moment of assignation is our disruptive awakening; it perpetually recurs and recalls ourselves to it in the spontaneous other-regarding phenomena that Danish theologian K. E. Løgstrup has called the "sovereign expressions of life,"[65] those moments of intersubjective melding that seem to us, when we are in their grip, to be the most obvious, perhaps the only, truths we can know. Without maintaining an openness to these sovereign expressions of life, we lose the motivating force for seeking to regard the other at all. Without allowing ourselves to be shot through in this sense by the possibility of others' suffering, ethics becomes susceptible to simple thematization, politics to rationalization, and even the most horrific neglect and abuse to normalization. Regard for the other becomes a matter of finding the "correct" line of moral considerability—the line that cannot be rationally or empirically doubted—rather than a matter of openly experiencing the plea that cannot be denied.

At this point, we can hear our economist friends urging—perhaps more strenuously than ever—the need to remain practical and to acknowledge hard choices about how to allocate inevitably limited resources. We should listen to these others also: There *is* a real danger that talk of infinite responsibility, of the limitless ethical demand, of the unbridgeable transcendence of the other, will cause us to lose touch with the immediacy of what we do know, of the innumerable victims of suffering that strain for recognition before us right now. If we simply take our foie gras with a reading from Levinas, we will succeed only in outsourcing responsibility to intellectual pursuits that, however profound and lyrical, lack meaningful connection to the flesh of the world. Instead, as Wood noted, "the other to whom one has an infinite asymmetrical

obligation must be thought of . . . in terms of the kinds of further relation that might be possible with that being, or else it becomes a purely
abstract relation."[66] Importantly, though, this need to concede a limited
role for ontology in structuring our ethical relations does not suggest
that we should embrace the economist's language of inevitable tradeoffs,
in which everything becomes a tradeoff for something else simply
through the positing of all value as monistic and exchangeable. The cor-
nerstone of ethics is not the economists' truth that tradeoffs are inevitable, but the more ancient truth, recovered brilliantly by Levinas, that
"[o]ne can exchange everything between beings except existing."[67] Welfare
economics dishonors this truth by assuming the assimilability of even
existence into its calculus of choice; as such, nothing is left unrational-
ized within cost-benefit analysis to signify the irredeemable loss of a life.
The ethical relation under welfare economics must be decomposed into
either an identifiable interest of the needy or an altruistic preference of
the fortunate. This approach is doubly mistaken, for the interests of the
needy are unknowable and the fortunate are already obliged, even before
they prefer.

To honestly recognize environmental law's others—to recognize
their faces *as faces*—we do well to support policies that give them flesh,
that attribute certain mortal needs to them, so that we no longer can comfortably maintain our effacing assumption that they may be treated as
just another element of the complex causal order, just another object to
be valued in a grand auction of survival. Absolute alterity *is* a reason for
denying that a program of action ever can truly honor others' existence,
meet their needs, or vindicate their rights, but it is not an excuse for failing to implement those programs that we believe will help them breathe,
drink, eat, and—if we continue to strive—flourish.

In short, we should step back, and then step up.

PART FOUR OUR ENVIRONMENTAL FUTURE

CHAPTER EIGHT

Ecological Rationality

THE IDEAL OF OPENNESS and responsiveness to an unknowable other does not readily lend itself to behavioral prescription; indeed, the very aim of ethics as first philosophy is to refuse to yield programmatic advice regarding how to live up to the infinite responsibility implied by the existence of other life, since all such advice inevitably does violence. Nevertheless, the critical point to take away from the previous chapter is that, although we will never know how to satisfy the infinite ethical demand of the unknowable other, *we do know how to fail it*. We fail the demand when we deliberately refuse to promote conditions that, in our best understanding, are necessary to support life. Accordingly, we should strive at present to ensure sustainable management of soil, forests, freshwater, and other natural resources, and we should seek to minimize destruction of the ozone layer, accumulation of greenhouse gases, and other actions that we believe will harm life. We should do these things not because our cost-benefit analysis demonstrates them to be optimal after having been properly stripped of conceptual prejudice, nor because our Kantian-inspired ethics posits them to be necessary to protect a newly expanded Kingdom of Ends, nor, finally, because our political theory identifies them to be the choice of hypothetical contracting agents, once those agents have been placed behind a veil of ignorance sufficiently opaque to mask whether they are called "man" or "cat" by

[handwritten margin note: Know how to fail an ethical demand]

man. Instead, we should do this because we have experienced what Levinas calls the "epiphany of the face," the recognition of a primordial responsibility for the mortality of the other, which is the beginning of all subjectivity and, by extension, all ethics and all politics.[1]

This chapter seeks to make these rather esoteric thoughts more tangible by examining a recent controversy over how best to regulate the withdrawal of cooling water from adjacent waterways by large power plants in the United States. Admittedly, the shift to this discussion from chapter 7 will be a jarring one, moving from the most abstract of continental theorizing to the most mundane of bureaucratic fish-counting. Yet the exercise is essential. Throughout this book, discussion has moved freely between theory and practice, between the high church of political liberalism and the low church of liberal politics, between the sophisticated utilitarianism of Mill, Bentham, and Sidgwick and the workaday welfarism of OMB officials. The implicit methodological premise has been that one cannot reliably cordon off these various levels of analysis from each other. Nor would such a cordoning be desirable even if it were possible. As Alfred North Whitehead reminds us, we must continually "recur to the concrete" in order to ensure that our abstract theorizing is informed both by what we are learning about the world and by what the world is doing with our learning.[2] In that respect, the gap between theory and practice in welfare economics is alarming, not only because actual policy decisions are being premised on inferior versions of welfare economics but also because they are being presented with a veneer of objectivity and precision that scarcely applies even to the physical sciences, let alone to a field so marbled with moral implications as welfare economics. Thus, although quite different in tone, the concrete analysis of this chapter aims to complement the philosophical musings of chapter 7; specifically, it aims to show that the questions at the heart of those musings are present in *every* environmental rulemaking, no matter how narrow or technical it may appear to be.

*

Electricity generating plants and other industrial facilities often depend on the withdrawal of water from rivers, lakes, and other waterways in order to manage excess heat generated during their production processes. The amount of water required is vast, on the order of 50 million gallons or more per day for a large plant and totaling billions of gallons of water per day for all facilities in the United States. The resulting environmental impact is also dramatic: The EPA estimates that more than 3.4 billion fish and shellfish (expressed as "age 1 equivalents") are killed by cooling water intake operations each year, either from being trapped against components of the cooling water intake structure and therefore suffering "exhaustion, starvation, asphyxiation, and descaling," or from being drawn into the cooling water system and therefore suffering "physical impacts," "pressure changes," "sheer stress," "thermal shock," and "chemical toxic effects."[3] These two mortality threats, referred to as "impingement" and "entrainment," affect not only the various fish and shellfish species for which the EPA was able to generate quantitative estimates during a recent rulemaking, but also certain endangered, threatened, and other "special status" species, such as sea turtles, Chinook salmon, and steelhead trout, as well as immeasurable quantities of phytoplankton and zooplankton that lie at the base of aquatic food chains. For all of these losses, regulators can only generate rough predictions of their likelihood and impact, given that "[p]opulation dynamics and the physical, chemical, and biological processes of ecosystems are extremely complex."[4]

impingement and entrainment

Cognizant of these kinds of informational difficulties, Congress, in section 316(b) of the CWA, mandated that "the location, design, construction, and capacity of cooling water intake structures reflect the best technology available for minimizing adverse impact."[5] The EPA's first effort to implement this best-available-technology provision was remanded on procedural grounds following an industry challenge in 1977, after which the agency formally withdrew the regulation in 1979.[6] In a hopeful gesture, the EPA reserved space in the Code of Federal Regulations for future cooling water intake rules.[7] Nevertheless, amidst the changing political climate of the 1980s and the agency's enormous backlog of regulatory responsibilities under other federal environmental statutes, the EPA abandoned its section 316(b) rulemaking efforts, leaving regulation of cooling water intake structures instead to the case-by-case decision making of pollution-discharge-permit issuers.[8] Eventually, facing a legal challenge by environmental groups, the EPA agreed in 1995 to a consent decree that required the agency to establish cooling water intake rules in multiple phases. Phase I, involving new facilities, was completed by the agency on December 18, 2001, and generally required

316 (b) CWA

Phase I new existing plants ↓

covered facilities to achieve environmental performance standards based on what is known as "closed-cycle cooling technology," a process in which cooling water is recycled and only periodically replenished from neighboring waterways rather than continuously withdrawn. Although environmentalists had argued on behalf of "dry cooling technology," an even more stringent approach that did not require the withdrawal of water at all, the Second Circuit in 2004 accepted the EPA's conclusion that the expense of this technology rendered it not reasonably "available" to industry.[9]

Phase II involved the much more politically nettlesome category of large existing power plants. The EPA's final regulations for this phase were issued on July 9, 2004, and offered a complicated array of compliance options, the most important of which were built around certain impingement and entrainment performance standards that had been identified by the EPA. These performance standards were particularly notable because they marked a refusal by the EPA to use closed-cycle cooling technology as the benchmark against which other proposed protection measures were to be evaluated. Despite acknowledging that impingement and entrainment provide the "primary and distinct types of harmful impacts associated with the use of cooling water intake structures,"[10] and that "closed-cycle, recirculating cooling [towers] . . . can reduce mortality from impingement by up to 98 percent and entrainment by up to 98 percent,"[11] the EPA nevertheless adopted standards that were based on the performance of less environmentally effective technologies. It did so because it chose to "interpret[] CWA section 316(b) as authorizing EPA to consider not only technologies but also their effects on and benefits to the water from which the cooling water is withdrawn."[12] Thus, the agency developed performance standards that required not the best technology available for minimizing impingement and entrainment, but rather the technology that best equalizes the marginal ecological benefits of reducing impingement and entrainment with the marginal economic costs of doing so.[13] This efficiency-oriented approach had a dramatic effect: The EPA estimated in its regulatory impact analysis that 125 facilities would adopt no impingement and entrainment controls at all under the Phase II rules.[14] Moreover, rather than up to 98 percent reduction in impingement and entrainment, most facilities would achieve only between 30.9 and 59.0 percent reduction in impingement and between 16.4 and 47.9 percent reduction in entrainment.[15] In short, the "best technology available for minimizing adverse environmental impact" became the technology that produced merely an acceptable cost-benefit ratio, irrespective of its overall level of environmental benefit.

The Phase II rulemaking provides a valuable opportunity to assess, in context, many of the arguments offered in previous chapters regarding the competing merits of cost-benefit analysis and more traditional approaches to environmental law and policy. Due to the EPA's unprecedented approach in the Phase II rulemaking, scholars now have available for study both an expansive cost-benefit analysis and the technology-based standard that would have been adopted by the EPA had the agency followed the straightforward technology-based approach that it took in the Phase I rulemaking. As this chapter argues, the EPA's Phase II rulemaking illustrates several limitations of cost-benefit analysis that have yet to be overcome in practice and that, in some cases, cannot be overcome even in principle. Most notably, the practical challenges that prompted Congress to amend the CWA in 1972, jettisoning an earlier attempt to correlate regulatory impositions with demonstrated ecological benefits, remain just as insurmountable today. Although many advances have been made in the understanding of ecosystem functioning and in methods of monetizing environmental impacts, the project of reducing environmental policy making to empirical technique remains deeply flawed. To be clear, no one denies the wisdom of acquiring information regarding the consequences of regulation as part of the policymaking process. But the wisdom of hinging regulatory outcomes on how well that information can be made to fit the form of a cost-benefit-analysis exercise has not been demonstrated. Instead, when viewed within the larger institutional context within which regulatory cost-benefit analysis must, by definition, occur, the approach appears to be an impractical and often illegitimate attempt to enshrine within law a particular political vision, one so deeply subscribed to by its adherents that they fail even to recognize its normativity.

[margin annotation: fail to recognize normativity of cost/benefit test]

In contrast, the simple heuristic contained within technology-based standards and other traditional precautionary approaches—in essence, "Do the best you can"—expresses great collective commitment to the preservation of human life and the environment without requiring satisfaction of unattainable informational demands by regulators.[16] Such seemingly "primitive" regulatory approaches can, under circumstances common to environmental, health, and safety regulation, reflect a greater degree of pragmatic wisdom than the elaborate technicality of cost-benefit analysis. In that sense, they can be described as "ecologically rational,"[17] a term developed by cognitive psychologists to describe adaptively sensible heuristics for situations in which optimality cannot be obtained. Because technology-based standards refuse to admit of a level of human or environmental harm that is optimal, they also are sensitive to the challenge of maintaining political commitment in the face of the unsatisfiable ethical demand posed by life's fragility. As explained below, the best-available-technology requirement of section 316(b), like other seemingly "irrational" or "extremist" environmental statutes, embodies a significant expressive wisdom in that it refuses to deny the irreplaceability of life or to embrace the utilitarian notion that an "optimal" amount of death ever can be identified.

*

Many of the shortcomings of cost-benefit analysis were inadvertently highlighted during the public notice and comment periods of the Phase II rulemaking by an influential environmental economist, Robert Stavins. In his public comments, Stavins attempted to take the EPA to task because, in his words, "the economic analysis offered by EPA in support of [its original proposed] rule [was] severely flawed, biased, and misleading."[18] In the economist's view, a series of "egregious errors" in the EPA's attempt to monetize the benefits of its technology-based standard made "a complete sham of the very process in which the proposed 316(b) rule [was] being considered."[19] Not only did these errors supposedly impart "a horrendous bias" to the EPA's results, supporting far greater levels of environmental protection than would be justified by "correct conceptual frameworks for economic analysis," but they also seemed to offend Stavins personally: He concluded his letter by stressing how "disappointing, troubling, and ultimately painful" it was for him to "review this analysis and provide these comments" in light of the "considerable time and effort over the past decade" that he had spent trying to teach the EPA staff more reliable economic methods.[20]

The EPA's purportedly "bogus"[21] valuation techniques actually represented innovative and context-sensitive efforts to grapple with the

OMB's regulatory-impact-analysis mandate within a statutory frame-work that largely rejects the narrow welfarist worldview embodied in conventional cost-benefit analysis. The choice of phrasing that Congress used in setting out the requirements of section 316(b) of the CWA—"best technology available for minimizing adverse environmental impact"—was not incidental. The CWA includes a dizzying array of technology-related standards, each carrying a subtle but significantly distinct meaning. Section 301(b) of the CWA, for instance, requires the EPA to establish initial effluent limitations for existing sources based on the "best practicable control technology currently available" (BPT). By 1989, the EPA was to replace those standards with more stringent ones based on the "best available technology economically achievable." Section 306, in contrast, requires effluent limitations for new sources to be based on the "best available demonstrated control technology [offering] the great-est degree of effluent reduction." These varying standards carry significantly different implications for the permissibility of cost-benefit analysis. Under the initial BPT approach, regulatory standards could be premised on an explicit comparison of compliance costs to environmental benefits. For the setting of second-generation existing source standards and new source standards, however, the EPA generally is prohibited from engaging in such comparison.

If the language of section 316(b) had been interpreted consistently with the language of these more stringent standards, then the EPA's cooling water intake structure rules could not have been premised on cost-benefit balancing. Instead, the agency would have been required to focus simply and directly on the affordability of increasingly efficacious environmental control technologies, recognizing that Congress itself already had determined that the benefits of regulation are sufficiently vast and difficult to quantify that only the "best" control technology will suffice. Nevertheless, following interventions by the OMB, Stavins, and industry commenters, the EPA in its issuance of the final Phase II rules bootstrapped its way into a cost-benefit-analysis regime by somehow reading the word "practicable" into the CWA's cooling water intake structure provision.[22] This rather brazen effort by the EPA to transform the best-available-technology standard of section 316(b) into a BPT standard—an attempt that ultimately received the blessing of a confused Supreme Court majority[23]—had followed a public dialogue that was initiated by the agency regarding alternative analytical approaches to the identification of site-specific best available technologies. The most important of the EPA's proposed alternatives included three tests: a "wholly disproportionate cost test," whereby the technology alternative would be chosen that offers the most environmental protection without producing costs wholly disproportionate to gains; a "significantly greater cost to benefit test," whereby the alternative would be chosen that offers the most environmental protection without producing costs that are significantly greater than benefits; and a "benefits should justify costs test," whereby the alternative would be chosen that maximizes net social benefits. Of these alternatives, the first two represented variations on long-standing judicial interpretations of the best-available-technology test, in which the EPA is permitted to grant firms an escape hatch if the most protective technology identified would produce environmental benefits that are unequivocally swamped by compliance costs. The third alternative, on the other hand, represented a direct application of cost-benefit analysis.

From among these alternatives, Stavins naturally acknowledged only one viable candidate: "Mainstream economic thinking—as it is taught in any university from Maine to San Diego, from Miami to Seattle—points to one and only one of these alternative criteria as able to lead consistently to decisions that are in the general social interest . . . the so-called 'efficiency condition' in economics."[24] The other alternatives, in Stavins's view, did nothing to ensure that the EPA would respect the Kaldor-Hicks efficiency test and, thus, they "virtually guarantee[d] that social decisions will not be welfare-improving, indeed [they] guarantee[d] that selected actions will make the world considerably worse off."[25] Behind Stavins's contentions lies an assumption that the policy space within which the EPA operates is informationally rich and probabilistically sophisticated, meaning that the agency easily can identify courses of action that maximize expected welfare outcomes. Stavins devoted no meaningful discussion in his comments to the possibility that the scientific and economic information necessary to fulfill textbook efficiency analysis may be lacking or that nonoptimizing decision criteria may be more pragmatically sensible under conditions of epistemic uncertainty and administratively costly decision making. Hence, when the economist wrote that only the cost-benefit test "will—by definition—lead consistently to decisions which make the world better off,"[26] he failed to acknowledge that cost-benefit analysis only does so "by definition," that is, by an a priori assumption that the information needed to satisfy the form of the optimization exercise is both attainable and costless.

In the case of cooling water intake structures, this assumption is wholly unwarranted. In order to meet the form of regulatory calculus demanded of it by the OMB, the EPA reduced a complex and highly uncertain policy decision to a simple question of how much to invest in "reductions in impingement and entrainment as a quick, certain, and consistent metric for determining performance."[27] Increased fish survival became the primary determining factor in the EPA's analysis simply because, at least for those fish that are commercially or recreationally valuable, the factor offered an ecological benefit that was readily quantifiable and monetizable. As the agency acknowledged, however, the potential impact of cooling water intake structures is much broader and more complex than direct mortality effects (which are themselves only roughly estimable):

Complexity of impingement and entrainment impacts ↓

> In addition to their importance in providing food and other goods of direct use to humans, the organisms lost to [impingement and entrainment] are critical to the continued functioning of the ecosystems of which they are a part. Fish are essential for energy transfer in aquatic food webs, regulation of food web structure, nutrient cycling, maintenance of sediment processes, redistribution of bottom substrates, the regulation of carbon fluxes from water to the atmosphere, and the maintenance of aquatic biodiversity. Examples of ecological and public services disrupted by [impingement and entrainment] include:
>
> - decreased numbers of ecological keystone, rare, or sensitive species;
> - decreased numbers of popular species that are not fished, perhaps because the fishery is closed;
> - decreased numbers of special status (e.g., threatened or endangered) species;
> - increased numbers of exotic or disruptive species that compete well in the absence of species lost to [impingement and entrainment];

- disruption of ecological niches and ecological strategies used by aquatic species;
- disruption of organic carbon and nutrient transfer through the food web;
- disruption of energy transfer through the food web;
- decreased local biodiversity;
- disruption of predator-prey relationships;
- disruption of age class structures of species;
- disruption of natural succession processes;
- disruption of public uses other than fishing, such as diving, boating, and nature viewing; and
- disruption of public satisfaction with a healthy ecosystem.[28]

These various ecological and public benefits garnered no monetary estimation in the EPA's economic analysis. Indeed, as the agency candidly admitted, even its focus on impingement and entrainment losses was highly incomplete: The agency "was not able to monetize benefits for 98.2% of the age-one equivalent losses of all commercial, recreational, and forage species for the section 316(b) Phase II regulation."[29] Thus, only 1.8 percent of one category of environmental benefit was monetized.

98.2% of impacts

Even for the limited data on cooling water impacts that it did have available, the EPA warned that "[b]ecause of . . . methodological weaknesses, EPA believes that studies . . . should only be used to gauge the *relative* magnitude of impingement and entrainment losses."[30] Yet these studies provided the raw material for the agency's economic analysis that in turn rejected dry cooling and closed-cycle cooling technologies in favor of weaker performance ranges. Nor was the advice to use impingement and entrainment studies only to gauge "relative magnitude" the only unheeded disclaimer in the rulemaking record. The agency also at various points warned that "[t]o rely only on estimated use values would substantially undervalue the benefits of the final section 316(b) rule"; that "[t]he organisms that remain unvalued in the analysis provide many important ecological services that do not translate into direct human use"; and that "[t]o the extent that the latter are not captured in the benefits analyses, total benefits are underestimated."[31] Elsewhere, the agency offered the sage advice that "[a] comparison of complete costs and incomplete benefits does not provide an accurate picture of net benefits to society,"[32] and that "there is a real possibility that ignoring non-use values could result in serious misallocation of resources."[33]

Despite these multiple and seemingly sincere disclaimers, the EPA ultimately could not resist claiming that its "proposed rule has the largest estimated net benefits, $452 million, of the five regulatory options analyzed."[34] Thus, one important level of objection to the cooling water intake rulemaking focuses on the agency's decision to allow cost-benefit analysis to heavily influence the ultimate selection of environmental performance standards when the analysis itself was woefully incomplete and uncertain. With so many positive effects of regulation remaining off the balance sheet, the agency actually had little reason to be confident that the conclusions offered by cost-benefit analysis were welfare maximizing. Nonetheless, as in other contexts, the promise of an "objective" quantitative analysis seemed difficult to resist in the face of a heavily politicized, deeply uncertain, and morally fraught decision. This cognitive lure was especially evident in Stavins's public comments. The economist wanted firmly to believe that regulators could identify the technology "which protects the target resources . . . up to the point where the incremental benefit from increased protection *just equals* the incremental cost of increased protection."[35] Because this standard of empirical sophistication is rarely even approached in the area of environmental regulation, important policy judgments must be made regarding how to handle information gaps, scientific uncertainties, system complexities, and other quantitatively intractable features of regulatory decisions. It is with respect to *these* judgments that welfare economists run the risk of confusing "personal opinion" with "professional expertise"[36]—the great failing that Stavins claimed to see in the public comments of a different commenter who disagreed with his optimific approach. It is also with respect to these judgments that the precautionary approach aims to serve a democratizing function in risk regulation, opening up for broader confrontation the unavoidable gaps, conflicts, and ambiguities of formalized policy making.[37]

These confusions over the role of economic analysis within policy making were also seen in a difference of view that arose between Stavins and environmental economist Frank Ackerman regarding whether "non-use" values—that is, values attributable merely to the fact of a species existing in the world, rather than being subjected to recreational, dietary, or other instrumental use by humans—should be estimated by the EPA. Beginning with the background precautionary assumption that not everything valuable about the environment can be discerned, dissected, and quantified by present human modes of understanding, Ackerman advocated the use of admittedly imperfect attempts to quantify and monetize the variety of ecological impacts that go beyond simple reductions in commercially and recreationally valuable fish mortality. For instance, citing a literature review of studies of use and non-use values in the environmental context, he noted that agencies could rely on a simple presumed ratio of two dollars of non-use benefit for every one dollar of use benefit in order to provide approximate numerical estimates of the former.[38] This ratio was offered in contrast to the EPA's customary "50 percent rule," which had been used to set non-use benefits at one-half of use benefits, but which Ackerman argued was based on outdated studies. Regardless of the ultimate heuristic chosen, however, Ackerman's most fundamental point was that the EPA should "avoid placing an effective value of zero on categories of value that the Agency does not have time or resources to analyze in detail."[39] Any other approach would be inconsistent with the environmentally precautionary aim of the CWA.

Stavins, in contrast, regarded the literature on use to non-use value ratios to be inadequate to support quantitative estimations for policy making. Because the studies underlying the literature review cited by Ackerman addressed a variety of environmental and natural resource issues other than the specific one of cooling water intake impacts, Stavins argued that the studies provide "no evidence of why, given the *specific* environmental improvements associated with the proposed regulations, non-use value should be of any *specific* magnitude."[40] By demanding original, detailed, and unambiguous valuation studies of non-use benefits and other ecological impacts, and by refusing to assign any non-zero value in the absence of such studies, Stavins's approach assigned an implicit burden of proof to the regulatory agency that predictably biases decisions against environmental protection. Such exactitude might be appropriate within the ivory towers of the university, where scholars aim to bolster the scientific credentials of welfare economics by portraying it as the "objective implementation of benefit-cost analysis, based on established economic theory and empirical research."[41] In the real world of policy making, however, decisions must be made in advance of comprehensive knowledge. Stavins nowhere defended his implicit assumption that this inevitable uncertainty must be construed against the environment—a surprising omission given that Congress, through the imposition of best-available-technology standards under the CWA, seemed to legislate a far more precautionary assumption.[42] Nevertheless, Stavins's comments achieved their desired effect: In light of the intense criticism that the EPA received even for its cautious use of a 50 percent ratio to estimate non-use benefits, the agency ultimately assigned no numerical value at all.[43]

why presume against env. re: uncertainty

A second, and potentially more significant, level of objection to the EPA's approach focuses on the opportunity cost of deploying regulatory cost-benefit analysis. As noted above, Stavins throughout his comments ignored one of the defining characteristics of environmental law, which is its demand that society make choices well in advance of thorough and reliable information regarding the consequences of those choices. Accordingly, the precautionary approach to environmental law and policy holds that regulatory decision making should not merely react to existing information, but also actively intervene in the processes and institutions by which information is generated. For instance, through careful assignment and management of the burden of proof, regulators may be able to marshal the information-generating resources of firms and other private actors in service of the public's environmental aspirations. By instead placing the burden on regulators to identify, quantify, and monetize potential adverse impacts of market activity, cost-benefit analysis forfeits a valuable opportunity to utilize incentives, penalty defaults, and other regulatory strategies directly in furtherance of informational goals. Such an opportunity is especially significant in the environmental regulatory context, given that, as public choice theory would predict, the regulated community is typically better represented and better resourced than nongovernmental organizations and other representatives of the public interest during administrative rulemaking.[44]

*

Even assuming that the EPA did somehow have reliable and comprehensive information regarding the myriad complex environmental impacts of cooling water intake processes, under cost-benefit analysis the agency still needed to transform those impacts into monetary values in

order to enable quantitative comparison with the expected costs of protection technologies. As noted above, when designing the Phase II rules,
the EPA focused only on "reductions in impingement and entrainment
as a quick, certain, and consistent metric for determining performance."[45]
Moreover, reductions in impingement and entrainment were valued
only insofar as they might result in identifiable gains to commercially
and recreationally valuable fish.[46] This choice of metric was useful for
the EPA since it allowed the agency to avoid intractable valuation questions that would have accompanied the effort to account for threatened
or endangered species impacted by cooling water intake processes.[47]
Indeed, the entire approach of focusing on reductions in impingement
and entrainment as "a convenient indicator of the efficacy of controls in
reducing environmental impact"[48] seemed designed to ensure that the
agency could appear to have satisfied the quantification and monetiza-
tion demands of cost-benefit analysis, irrespective of the actual share of
the consequentialist landscape that was encompassed by the agency's
calculations.

In earlier stages of its analysis, the EPA candidly acknowledged that
existing valuation techniques in the environmental economic literature
tend to understate the benefits of environmental protection by focusing
only on the most readily understood and quantifiable effects of human
interference with ecosystems. Thus, the agency sought to supplement
its initial analysis with an indirect measure of the value of environmental protection, one that asks what the cost would be of replacing the variety of goods and services that are provided by a healthy functioning
ecosystem. While conventional valuation techniques ask whether an environmental resource is worth saving based on estimates of the monetary worth of its fruits, the EPA's "habitat replacement cost" (HRC) method
instead asked what those fruits are worth based on how difficult and
costly it would be to develop substitutes for the environmental resource
that generates them. The former approach reflects a demand-side estimate of environmental value based on the amount that individuals appear willing to pay in order to preserve discrete environmental goods or
services; the latter approach generates a supply-side estimate of environmental value based on the amount that society would need to expend in
order to replace those same environmental goods and services in their
interrelated ecological context.

Stavins strongly condemned HRC as "a completely illegitimate method of analysis," stating that it is "essentially oxymoronic and completely invalid."[49] In Stavins's view, the HRC methodology commits "one of the gravest of errors in economics" by confusing environmental costs and benefits.[50] HRC estimations essentially assume that an object of environmental protection is a unique capital resource that produces a flow of valuable goods and services, the worth of which can only be approximated by asking what it would cost to develop a substitute resource that produced those same goods and services. To Stavins, this approach is wrongheaded, since the very point of regulatory cost-benefit analysis is to ask whether the environmental resource is worth preserving at all. That is, rather than just assume that society must have clean water, biodiversity, and the variety of other goods and services that flow from intact ecosystems (or their built replacements), Stavins argued that regulators must estimate the monetary amount that individuals are willing to pay in order to obtain those specific goods and services. In this view, no resource—not fish, not freshwater, not human life—is considered sufficiently important or irreplaceable to avoid being subjected to productive conversion.

This view recalls the more general theoretical distinction that exists between conventional environmental economists who tend to view all of natural and human-made capital as substitutable, and ecological economists, who view many natural resources and ecosystem services as practically irreplaceable. In this context at least, Congress has sided with the ecological economists: The goal of the CWA is to restore and maintain "the chemical, physical, and biological integrity of [the] Nation's waters."[51] The goal is not to view those waters as merely contingent resources to be impaired or sacrificed at any moment for the promotion of an abstract and undifferentiated maximization of welfare. Nor is the CWA at all unusual in this regard: As argued in earlier chapters, one way of understanding much of the project of environmental law is as an effort to identify elements of the environment—for instance, particular species, habitats, ecosystems, or global atmospheric processes—that are sufficiently important to human well-being or sufficiently worthy of admiration and respect in their own right to take them outside of the realm of instrumentalist trading.

Nonetheless, to Stavins, value remains a steadfastly monistic concept: "In economic terms, the benefits of some action are equivalent to

Stavins: Value is monistic

the aggregate of the willingness to pay . . . by the affected human populations for that action or outcome."[52] From this perspective, everything
of value in the world can be readily commensurated, since, by assumption, value only takes the form of individual human preference, as manifested in measurable expressions of willingness-to-pay. This tendency
toward value monism also was apparent in the EPA's flirtation with a
trading program as an additional compliance option for firms under the
final Phase II rules. Under this radically commensurated approach,
facilities would have been permitted to purchase credits from other
firms representing environmental impact reductions equivalent to the
purchasing facilities' regulatory obligation.[53] Interestingly, here the EPA
balked at the "comparability and implementation challenges" implied by
such a trading program: "EPA does not believe that it is possible at this
time to quantify with adequate certainty the potential effects on ecosystem function, community structure, biodiversity, and genetic diversity of
such trades, especially when threatened and/or endangered species are
present."[54] Those same challenges, however, also applied to the attempt
to conduct a cost-benefit analysis of cooling water intake structure requirements; thus, the EPA should have been equally hesitant to undertake *that* analysis.

   In addition to these practical concerns, the value-monist approach
of welfare economics also must meet the ecological economist's more
fundamental conceptual objection that willingness-to-pay valuation measures confuse the categories of income and capital, flow and stock, present and future. As noted in chapter 6, from the strict welfare economic
perspective, a fishery would only be managed sustainably if the anticipated benefits of doing so happen to justify the costs. If it turns out that
"liquidating" the fishery and reinvesting the monetary proceeds in other
investments offers a higher net present value than sustainable use, then
nothing in the value monism of welfare economics counsels against
such liquidation. Characterizing natural resources as capital goods or
stocks, on the other hand, illuminates their irreducibly intergenerational
aspect. As argued in chapter 6, the intuitively appealing criterion of maximum sustainable yield represents, at bottom, a normative judgment regarding intergenerational distributive equity: The value of a fishery's
yield is considered sufficiently large and distinctive to merit preserving

the underlying capital stock, so that the same annual flow continues ir-respective of the opportunity costs implied by sustainable management.

For the cooling water intake rulemaking, these fundamental concep-tual issues manifested in the question of whether the EPA underesti-mated the commercial and recreational value of fish by only calculating the value of fish that eventually are landed. From the ecological eco-nomic perspective, uncaught fish should be credited as valuable since they form the stock from which future fish grow, including those caught fish that later register in the EPA's benefit calculations. Treating natural capital instead as a pure consumable may lead to a situation that is deemed efficient even as it fails to ensure sustainability. Like welfare-maximization models that are able to promote welfare even to the point of rendering future humans extinct, the EPA's focus on only *individually* harvested fish ignores the health of the underlying communities from which those fish emerge. This conceptual issue should have played a far more prominent role in the EPA's decision making, given the dire state of the nation's fisheries. Even the EPA acknowledged that "some studies estimating the impact of impingement and entrainment on populations of key commercial or recreational fish have predicted substantial de-clines in population size."[55] Likewise, the value of threatened, endan-gered, and other "special status" species was left unquantified in the EPA's regulatory impact analysis, despite evidence that such species are threatened by the impacts of cooling water intake. Moreover, the at-tempts that the agency did undertake to value threatened and endan-gered species simply treated species preservation as an investment decision to be based on individuals' revealed preference for species sur-vival. From this perspective, certain charismatic species, such as the log-gerhead sea turtle, might fare well based on their high nonconsumptive value as objects of photography and other indirect uses, while more "ob-scure species" would be relegated to the notoriously difficult category to estimate of "pure non-use value."[56] Although preferable to the EPA's eventual failure to credit endangered and threatened species impacts at all, the investment approach nevertheless fails to take seriously the awe-someness of extinction as a collective deed. Like the irreversible deple-tion of a once-grand fish stock, species extinction is a deed that lies outside of purely secular time. Once these stocks and species are gone, they are gone. Thus, as chapter 6 explained, the challenge of establishing appropriate standards of protection is not one of maximizing allocative efficiency—of squeezing the most net present value out of resources that are assumed to belong to the current generation—but instead is one of determining whether, as a matter of intergenerational distributive equity, those resources should remain available for the benefit of future generations . . . or perhaps even for their own benefit.

*

Because the EPA eventually felt the need to abandon both its conservative 50 percent rule for non-use value estimation and the HRC measure of ecosystem value,[57] the agency ended up capturing only a very partial picture of the value of cooling water intake regulation. The agency did present what it styled as a "break-even analysis," which illustrated the amount of total and per capita annual valuations of missing ecological impacts that would be required in order for the rule's costs to equal its benefits. This approach—which essentially allowed policy makers and their various publics to examine the value of habitat preservation that is implied by the CWA's precautionary mandate and to ask themselves, "Is it worth it?"—bore some relationship to another valuation technique that the EPA was forced to back away from after harsh public comments, the "societal revealed-preference" method. Under this approach, the agency examined the level of compliance costs that had been absorbed by society in connection with previous governmental actions that sought to attain the kinds of ecological benefits that the section 316(b) rulemaking would promote. Like the agency's HRC and break-even analyses, the societal revealed-preference methodology represented an innovative way of responding to the OMB's demand for quantitative, monetized information regarding the costs and benefits of the Phase II rule, despite the great uncertainty characterizing ecological effects.

Stavins, however, was no more pleased with the societal revealed-preference methodology than he was with HRC: "Like the HRC method, this approach has no foundation in economic theory, is not accepted by economists as a legitimate empirical method of valuation, and is no more than a method of cost analysis mistakenly applied to the benefit side of the ledger."[58] Stavins took particular pains to emphasize that, in his view, the "proposed method is *not* a revealed-preference method."[59] Indeed, he complained that the societal revealed-preference concept is "a complete corruption of the notion of a *revealed-preference* method."[60] Through such comments, Stavins again insisted on a relentless value monism, chiding the EPA and commentators such as Ackerman for believing that "choices made by government can be used objectively, in the same manner as individual choice, to establish valid measures of benefits."[61]

This controversy over the societal revealed-preference methodology was instructive, since it raised in a discrete setting the basic normative question posed by Mark Sagoff in his classic critique of cost-benefit analysis: What, after all, is wrong with allowing valuations of public goods to emerge from society's willingness to act collectively as citizens to preserve those goods? What is wrong with simply allowing individuals to express their preferences for environmental goals directly through political channels?[62] To be sure, a monetary value of the environment will be implied after the fact by the level of resources that are necessary to fulfill adopted environmental policies—one that will even be quantitatively estimable through subsequent use of the societal revealed-preference methodology—but this value will not drive the initial selection of policy. Rather the monetary value will simply be an ancillary effect of a policy choice that was premised on social values, explicitly discussed and mediated through democratic decision-making processes.

Seen in this light, the most critical question raised by cost-benefit analysis is not to be found in the details of ever-more-refined modes of valuation, but rather in the initial staging decision regarding what is, and is not, a proper subject of instrumentalist trading via revealed-preference cost-benefit exercises. Perhaps a more powerful example of this distinction for the welfare economist would be the classical liberal principles of strong private property rights and freedom of contract: Should the content of *these* laws be determined through cost-benefit analysis? If so, how would the relevant valuations of cost and benefit be derived, since the very laws under consideration are the ones that give rise to the market structure and exchange activities that enable revealed-preference analysis? Like the basic liberal protection of property rights and freedom of contract, many of the central goals of environmental law and policy also are of a foundational character and therefore cannot adequately be posed within the language of cost-benefit analysis. Instead, they must be answered through the more familiar discourses and procedures of constitutional democracy, a point returned to in the next chapter.

Stavins's strong opposition to the HRC and societal revealed-preference methodologies seems to have been driven by a suspicion of precisely those discourses and procedures. After all, if he was willing to concede that group behavior can reveal meaningful benefit information,[63] why did he balk at the deliberations of the larger group that comprises a political community? Nor, conversely, is it clear why individual willingness-to-pay studies should be deemed the gold standard for welfare valuation. Stavins seemed to imply that the societal revealed-preference methodology is especially unreliable because there is no guarantee that individuals "*actually (and voluntarily) incur[] costs* to avert (or tolerate) the environmental disruption in question."[64] The very domain of cost-benefit analysis, however, is one in which private-market activity—individuals "*actually (and voluntarily) incurring costs*"—is presumed to have failed to maximize welfare. Indeed, revisions during the Bush Administration to the executive order mandating cost-benefit analysis purported to make the identification of some market failure along these lines a prerequisite to agency action.[65] By assumption, then, conventional willingness-to-pay valuation exercises must apply outside of the very context that is said to make them especially reliable as indicators of value. This point is oddly underappreciated: In his comments, for instance, Stavins devoted a great deal of critical attention to the practice of "benefits transfer analysis," whereby values derived from one economic study are transferred for use in a somewhat analogous, but nevertheless distinct, regulatory setting. What Stavins should have noted is that every willingness-to-pay study, when used in a regulatory context, is by definition a benefits transfer exercise entailing deep conceptual and practical difficulties.

For some observers, this steadfast attempt to force environmental policy decisions into the rubric of atomized, private-market decision making is simply misguided. Perhaps most famous among such observers is the Nobel laureate economist Amartya Sen, who has written that "[t]he very idea that I treat the prevention of an environmental damage just like buying a private good is itself quite absurd. . . . [I]t would be amazing if the payment I am ready to make to save nature is totally independent of what others are ready to pay for it, since it is specifically a social concern."[66] In his Phase II rulemaking comments, Stavins argued that this quotation from Sen merely reflects the commonplace notion that, in the case of public goods such as environmental conservation, individuals can be expected to free-ride, thereby driving a wedge between their true demand for public goods and their observed (unregulated) contributions to the provision of such goods. But in fact, Sen was making the much stronger claim that specifically social concerns demand specifically social methods of deliberation, valuation, and decision making. As he has noted in another context, "[w]e can have many reasons for our conservational efforts—not all of which are parasitic on our own living standards and some of which turn precisely on our sense of values and of fiduciary responsibility."[67]

Because of this value pluralism, which he believes is inherent in contemporary liberal societies, Sen has emphasized the need to guarantee to individuals the "freedom to participate" in environmental policy making. Thus, the famed economist and political theorist has cast his lot not with conventional revealed-preference methodologies or even with the supposedly radical economic methodologies of HRC and societal revealed preference, but rather with the kind of democratic forums, institutions, and processes championed by the likes of Sagoff: "The relevance of citizenship and of social participation is not just instrumental. They are integral parts of what we have reason to preserve. We have to combine the basic [instrumentalist] notion of sustainability . . . with a broader view of human beings—one that sees them as agents whose freedoms matter, not just as patients who are no more than their living standards."[68] Similarly, laws and policies should not be seen merely as instrumental tools for serving human "patients"; they should be recognized as repositories of cultural value and meaning that both flow from and help to form human "agents," including even the collective agency formed by a political community.

The temptation to reduce this complexity is great. The Supreme Court majority, for instance, determined that the best-available-technology standard of section 316(b) could be read to permit cost-benefit analysis by the EPA, since the agency had long rejected technologies whose costs are "wholly disproportionate" to environmental benefits.[69] Such limited consideration of costs, however, should not be taken to render best-available-technology standards extensionally equivalent to cost-benefit analysis. Technology-based standards continue to carry vastly different *expressive* connotations than cost-benefit analysis— connotations that are far more resonant with the ethical demand laid out in chapter 7. The determination by Congress that an environmental, health, or safety goal is sufficiently important that only society's collective "best" efforts will suffice opens an inevitable and lamentable gap between statutory aspiration and regulatory achievement. Accordingly, those harms that are not prevented under a best-available-technology standard serve as moral remainders, signifying a collective need to seek ways of doing "better" in the future, of further protecting life and lowering environmental impact. Under cost-benefit analysis, those harms that are not avoided simply represent the "right," the "efficient," or the "optimal" level of harm. Indeed, precisely because it purports to account for all relevant consequences, cost-benefit analysis must round to zero the moral remainders of risk regulation, leaving nothing further to signify a societal need for continuing vigilance. In the EPA's Phase II rulemaking, for instance, literally billions of fish each year were simply ignored by the agency's economic analysis, treated as if their loss was meaningless because the question of their worth had been abandoned.

*

The final lesson of the Phase II rulemaking pertains to the role of academic theory in law and politics. Like many welfare economists, Stavins claimed to operate within a methodological system in which so-called "normative" questions can be kept separate from so-called "positive" questions. When evaluating the EPA's initial proposal to require closed-cycle cooling technology, for instance, Stavins was confident that the EPA's analysis rested on "flawed reasoning, confused concepts, and fundamentally invalid research methods."[70] He was sure that if only the agency would apply proper welfare-economic techniques, it would come to appreciate the "tremendous inefficiencies that exist in regard to the targets that are established in the proposed rule."[71] With respect to the more ecologically informed comments of Ackerman, Stavins was equally confident that his fellow economist had committed the unforgivable sin of "confound[ing] objective questions about the implementation of cost-benefit analysis with normative questions about its use in decision-making."[72] Stavins did generously allow that "[p]hilosophically or ethically based opposition to the use of benefit-cost analysis surely ought not to disqualify someone from commenting on a proposed rule."[73] Nevertheless, he insisted that such "philosophical" or "ethical" discussion must be kept separate from "questions about the correct measurement of benefits and costs,"[74] and that "normative and positive dimensions should not be intermixed and confused."[75]

Stavins failed to perceive, however, that numerous, inescapably ethical and political judgments go into constructing the supposedly settled distinction between normative and positive dimensions of welfare analysis. Moreover, he failed to perceive that those judgments also are directly at issue in much of environmental law and policy. Thus, even if Stavins is correct that he accurately represented the mainstream of the welfare-economics profession, he provided no reason for either the EPA or the world outside of economics to accept the profession's self-understanding of its intellectual project. Put differently, it is only welfare economists who believe that the foundational ethical and political judgments at the heart of cost-benefit analysis are sufficiently settled that they can be bracketed into a separate and generally ignored domain of "normative" analysis. In fact, numerous times in the history of modern environmental lawmaking, including arguably in section 316(b) of the CWA, the welfare-economic approach to issues of pollution control, resource management, and other environmental problems has been expressly and roundly rejected because of collective disagreement with one or more of its foundational assumptions.

When applied outside the context of disciplinary self-regulation within a university environment, the economist's insistence on separating "positive" and "normative" analyses becomes tantamount to an effort to extend the profession's preferred values into other spheres of life. Indeed, when one observes how sloppy the profession can be in handling its own purportedly settled distinction, one begins to wonder whether even ardent proponents believe that they are obtaining "correct measurement of benefits and costs." This is made all too plain in Stavins's description of welfare economics, in which he attempts to establish the discipline's scientific bona fides and to rebuke those who, like Ackerman, would call them into doubt through their support of heterodox methods:

> In contributing to public policy debates, the economics profession has long since established a clear distinction between "positive" and "normative" economics. While the former deals with questions for which objective and rigorously proven answers can be provided, the latter deals with questions that inevitably involve opinions informed by individual values. The proper *implementation* of benefit-cost analysis falls squarely within the arena of positive economics. Rigorously established methods that are firmly rooted in widely-accepted economic theory can provide objective information on whether a regulation is economically efficient, that is, whether the amount that all those benefiting from a regulation would be willing to pay for those benefits exceeds the total amount that would need to be paid to individuals burdened by the regulation to ensure that no one is made worse off.[76]

In this passage—in the very context of defending the "objective" credentials of positive economics—Stavins commits "one of the gravest errors in economics," that of confusing willingness-to-pay and willingness-to-accept valuation criteria. The choice between these two valuation approaches can radically affect cost-benefit-analysis results, since the choice essentially reduces to a normative judgment regarding which actor should be initially endowed with a legal entitlement. Valuations may differ dramatically depending on whether one must pay to obtain an entitlement or instead can demand an amount of monetary compensation adequate to compel voluntary relinquishment of the entitlement.[77]

Welfare economists hope to avoid such overtly normative questions, and thus they typically adopt willingness-to-pay valuations without much explanation. When pressed, they typically offer a practical argument that willingness-to-pay valuations are better "behaved" than willingness-to-accept valuations, since the former are disciplined by actual or hypothetically imposed budget constraints and therefore tend to give more consistent (and lower) values. As Stavins noted, the conventional privileging of revealed-preference studies over expressed-preference or contingent valuation studies is often defended on similar pragmatic grounds.[78] But these choices also have the convenient normative effect of forcing workers, consumers, fish, and other imperiled entities to demand greater levels of protection only through markets, using their existing rights and resources, rather than through law. The transformative potential of law is thus constrained because valuation is tethered to the status quo—the same problematic status quo that gave rise to a need for law. We should hesitate to view the more-stable behavior of willingness-to-pay valuations as evidence that society can identify the "optimal" level of death through conventional cost-benefit techniques. Instead, we should see those consistently lower willingness-to-pay valuations as evidence that workers, consumers, and fish ought to be given more power directly through law, in order to help close the gap between the two valuations.

Stavins well knows the difference between willingness-to-pay and willingness-to-accept valuation frames. As he stated elsewhere in his comments, "[i]n economic terms, the benefits of some action are equivalent to the aggregate willingness to pay . . . by the affected human population for that action or outcome."[79] Thus, his conflation of the two criteria when defending the rigor of welfare economics as an academic discipline is curious. Perhaps the explanation for his apparent slip lies in the superior rhetorical force of the willingness-to-accept valuation terminology. Normatively, an outcome seems much more defensible if the winners benefit by an amount exceeding the loser's loss, measured not just by how much the losers would be willing to pay to avoid the loss, but by how much they would "need to be paid . . . to ensure that no one is made worse off," as Stavins put it. The latter characterization seems to suggest that the losers lose not because they already *were* losers, endowed with inadequate resources to back up their preferences sufficiently to alter market outcomes, but instead because they themselves, when given a meaningful choice, determined that society is better off for their suffering.

By invoking the rhetoric of willingness-to-accept valuations in a context where it does not apply, Stavins claimed for welfare economics a degree of normative deference that the discipline does not deserve. Conversely, by ignoring the many roles of law that extend beyond self-surrender to the social welfare function, he understated the wisdom of more traditional precautionary approaches to environmental law and policy.

Normative vs. Positive framing is key...

Are there any values here which are in tension?

CHAPTER NINE

Environmental Constitutionalism

THE UNITED STATES CONSTITUTION is one of the few such texts in the world that fails to explicitly address environmental protection.[1] Quite to the contrary, the U.S. Constitution contains a variety of features—including limited enumeration of national legislative powers, strong protection of private property rights, judicial scrutiny of interstate commerce regulation, and imposition of standing and other justiciability requirements—that arguably restrict the country's ability to address the environmental needs of present and future citizens.[2] And while it is true that many U.S. state constitutions specifically address concerns of pollution and resource conservation,[3] courts typically refuse to give such provisions self-executing force; instead, citizens are directed back to the same political institutions whose amenability to capture and presentism may well have provided the impetus for enshrining environmental protection within the state constitution in the first place.[4]

In light of this situation, scholars and activists have for years advocated the constitutionalization of environmental protection at the national level, whether via judicial interpretation of existing constitutional provisions or via formal amendment.[5] They recognize the challenges facing this project: Expanding the reach of constitutional protection to encompass future generations or nonhuman life-forms—not to mention the environment as such—would require significant adjustment to

229

the anthropocentric, individualistic, and instrumentalist outlook upon which liberal ordering strongly depends. Many environmental law scholars seek such adjustments immediately, for they see the Enlightenment ethos as an impediment not only to the ultimate achievement of cosmopolitan, intergenerational, and interspecies justice, but also to the successful resolution of more immediate environmental problems. Accordingly, they urge scholars to regain the spirit of intellectual chutzpah that once led the field to promote legal standing for trees, revival and expansion of the ancient public-trust doctrine, guardianship obligations on behalf of future generations, and other such revolutionary devices of recognition.[6] They also advocate tempering the heavy reliance within environmental policy making on science, economics, and means-ends rationality, for they believe that such an approach forces environmentalists to couch their goals within a framework that inevitably understates the case for sustainability.[7] Instead, they would restore the romantic, even illiberal spirit that once animated environmentalism, before it seemed to evolve into a technically oriented advocacy movement.

Other commentators, however, view with skepticism the effort to craft a post-Enlightenment environmentalism out of the existing legal order, believing that environmentalists instead should continue to assert a primarily welfarist agenda.[8] In their view, it may well be that the final goal for environmentalism *is* to become diffused within the generative grammar of society, so that sustainability simply emerges from the constitutional system as a matter of course, rather than being ever-contingent on choice and circumstance. Presently, however, skeptics of environmental constitutionalism contend that the aim to guarantee "certain basic, absolute levels of environmental protection" reflects a "combination of political idealism and scientific naivety."[9] As support for this claim, they note that several early federal environmental statutes already purport to impose absolute standards of protection, but the United States has lacked the resolve necessary to fulfill the statutes' literal terms. In their view, the central insistence of welfare economics—that tradeoffs are inevitable—dooms to failure any effort to reliably and stably identify safe minimum standards, absolute rights of environmental protection, or other such deontologically inflected notions. At the least, they believe that "constitutional enshrinement of any particular environmental policies seems premature," given the substantial "disagreement [that] remains over the socially appropriate levels and types of environmental protection."[10]

Perhaps this unresolved disagreement also explains why, for those countries that *do* have express environmental provisions in their constitutions, the provisions tend to be vaguely specified and weakly enforced, much like their counterparts within U.S. state constitutions. To be sure, there exist rare moments of constitutional specificity, such as the requirement of the constitution of the Kingdom of Bhutan that "a minimum of sixty percent of Bhutan's total land shall be maintained under forest cover for all time,"[11] just as there are occasional instances of serious legislative commitment to the incremental realization of environmentally based human rights, such as the program in South Africa to give concrete expression to the right to water for subsistence needs.[12] On the whole, however, efforts to constitutionalize environmental law remain largely symbolic exercises even under the socially and environmentally progressive constitutions that have been adopted during the past half century. The situation is comparably dim at the supranational level: As recent commentators conclude, "while there appears to be a growing trend favoring a human right to a clean and healthy environment—involving the balancing of social, economic, health, and environmental factors—international bodies, nations, and states have yet to articulate a sufficiently clear legal test or framework so as to ensure consistent, protective application and enforcement of such a right."[13]

largely symbolic exercises...

This final chapter seeks to reframe the question of environmental constitutionalism in light of this pessimistic backdrop. It sketches an approach to environmental constitutionalism that is incremental and pragmatic in pitch, yet that still might radically shift our understanding in the manner desired by modern environmental law's early and daring proponents. The chapter begins by arguing that the our failure to achieve consensus on how society should "trade off" economic and environmental interests is in substantial part an artifact of the framework itself for identifying tradeoffs. Because the problem of tradeoffs is posed in a language that constrains and prefigures our potential responses to it, the problem is never broached with the sense of solemnity, humility, and possibility that it deserves. As will be argued, it is precisely the need to decide under conditions of undecidability that gives environmental law much of its normative traction. Acknowledging that environmental law is, in that sense, impossible—that its goals will never be achieved and that our values will never be optimally balanced—is the very precondition for making environmental law possible. Accordingly, to improve our chance of agreeing on foundational norms of environmental protection, we must put the economic approach to environmental protection back in its place. We must never let its lesson obscure more ancient wisdoms, such as those embodied in the precautionary principle, which have been wrongly understood as competitors to cost-benefit analysis, rather than as simpler and more essential reminders of collective responsibility. As chapter 2 explained, those reminders must be heeded in order to motivate environmental lawmaking of any sort, including even the optimizing logic of welfare economics.

This chapter then draws from Bruce Ackerman's expanded understanding of constitutional development to argue that much of the vision necessary to achieve an environmental constitutionalism was in fact already written into the framework of modern environmental law, before that vision was untimely forgotten. Features of 1970s-era environmental law that are maligned from the perspective of welfare economics can in fact be understood as efforts to alter the underlying market substrate against which welfare economics seeks to optimize.[14] Today those efforts appear to be "inefficient" or otherwise misguided largely because their aim has been misapprehended: When we evaluate efforts to achieve a constitutional reordering through a lens that presumes constitutional inalterability, we confuse first- and second-order subjects. Remembering traditional environmental law as a first-order subject—as a constitutional aspirant—sweeps away the basis for much of the confusion and divisiveness that today plagues the field. Accordingly, rather than issue a call for a new constitutional convention or a laborious process of formal constitutional amendment, this chapter instead asks us simply to live up to the laws we already have, in the hope that the popular constitutional movement archived within those laws can be reawakened. As noted in this book's introduction, the frequent refrain that "It takes a theory to beat a theory" overlooks the critical fact that we already had a theory—and still do. Likewise, the frequent demand that environmentalists offer a comprehensive reform proposal for environmental law and policy ignores the fact that the existing framework contains promise yet unrealized.

Recognizing, however, that an expressively powerful jolt may be necessary to revive our environmental constitutionalist movement, the book closes with an appended statutory proposal, the Environmental Possibilities Act. Modeled after the infamous "supermandate" proposal of the 1990s that would have subjected all of federal environmental law to a cost-benefit test,[15] the Environmental Possibilities Act provides that no federal environmental, health, and safety law shall be construed to allow monetized cost-benefit balancing to *satisfy* its demands. Instead, under the act, those laws must always be interpreted and implemented in ways that expressly recognize the loss of life as irredeemable and that accept the need to constantly find ways of doing better by those who depend on us. Cost-benefit analysis would not be prohibited under the act, but rather resituated, in such a way that its seemingly optimific results would never obscure the importance of striving to hold open environmental possibilities. With its central focus on opportunity costs, welfare economics regards possibility as a measure of value to be assessed only when it is already closed, rather than as a prospect for adventure with the future to be held open at all cost.[16] Closure of possibility may indeed give a price for what instead was chosen, but it never does only this.

*

In Montana, unlike in most U.S. states, courts have interpreted the constitutional right of environmental protection broadly. As Buzz Thompson notes, the Montana provision is unique in several respects. It is the only U.S. state provision that deems the right to a healthy environment "inalienable"; it expressly includes "future generations" within its orbit of protection; it imposes an affirmative duty on citizens to protect the environment; and it provides for a detailed set of standards to both preserve and restore a "healthful" and "clean" environment.[17] Even more surprisingly, in the landmark case of *Montana Environmental Information Center (MEIC) v. Department of Environmental Quality*,[18] the Montana Supreme Court gave the constitutional provisions immediate teeth by conferring standing on private plaintiffs to challenge degradations of environmental quality and by stating a strict standard of review for courts to apply. As part of its analysis, the MEIC court determined that the Montana State Constitution embodies the precautionary principle, notwithstanding the onslaught of scholarly criticism that the principle had garnered.[19] And in case there were any remaining doubt about the court's commitment, a subsequent opinion held that the constitutional provisions could be used to challenge the activities of private parties, thereby opening the way to a gradual reassessment of the meaning and value of economic liberty in Montana.[20]

The significance of these holdings can be illustrated by a lower-court case involving a claim by residents of Sunburst, Montana, that their land and groundwater had been contaminated by pollution from a gasoline refining facility owned by Texaco. A jury found in favor of the aggrieved landowners and awarded damages equal to the estimated cost of restoring their property to its previous condition. Noting that earlier state precedents had capped damages at the market value of the property prior to the infliction of harm, Texaco argued that it should not be held liable for restoration costs in excess of market value and that the plaintiff's property instead should be considered permanently damaged. The court refused to cap damages in this manner, reasoning that when plaintiffs have suffered pollution, "[l]imiting those plaintiffs to recovery of diminished property value would do nothing to correct the environmental insult at issue."[21] Citing the state constitution, the court observed that "Montana's strong public policy of environmental preservation is furthered by allowing recovery of restoration damages which will be used to remove pollution from the environment."[22] In short, unlike the welfare-economic paradigm, which draws no distinction between prevention and compensation or between environmental harms caused and market opportunities forgone the constitutional culture of the state of Montana appears to be evolving toward a system of foundationalist environmental protection.

It is easy to underestimate how powerful such distinctions can be, once they are dispersed throughout the normative order of a complex sociolegal system and subjected to its potentially reinforcing and amplifying effects. It also is easy to underestimate how dramatically the tenets of welfare economics conflict with such a program of environmental constitutionalism. As an illustration, consider a recent cost-benefit assessment conducted by the EPA as part of its long-delayed effort to more stringently regulate mercury emissions.[23] Because many of the health benefits of mercury emissions reduction had been arbitrarily assigned to a separate air pollution regulation issued concurrently with the mercury rule—with both regulations eventually being overturned for having exceeded the EPA's authority under the CAA[24]—the primary health benefit assigned to the mercury rule consisted only of avoided instances of mental impairment from neonatal mercury exposure. To assess this impact, the EPA's regulatory analysts estimated the number of intelligence quotient (IQ) points that otherwise would be lost from mercury emissions, which the analysts transformed into dollar values by referencing empirical studies of the relationship between IQ levels and earning potential. Citing a need to be comprehensive, however, the analysts then *reduced* the dollar value of avoided IQ losses due to the fewer years of education that individuals with lower IQ levels typically seek (which, in turn, "saves" society the resources that would be lost through tuition expenditures and forgone years of productive labor).[25] Thus, the benefit of avoiding mercury-induced mental impairment was first reduced to a crude measure of the wage impact of intelligence and then further reduced to reflect the offsetting "costs" associated with higher IQ levels.

To return to the example of genetic engineering from chapter 7, suppose that the question facing society was not whether to allow mercury emissions to inflict harms on unborn children, but whether to directly engineer newborns with reduced intelligence. Some might invite philosophical debate over whether ignorance truly is bliss,[26] but would anyone seriously contend that the question should be resolved through monetary comparison of the child's earning potential with avoided tuition expenses? To see why the EPA's approach was so inadequate, consider what would have followed if the opportunity cost of higher intelligence had turned out to be *greater* than the enhanced earning potential. If it is true that "everything depends on what the facts turn out to be," as the cost-benefit optimizer would have it, then it could well have been optimal for the EPA to mandate *increased* mercury emissions, since the IQ level of our children would have appeared to be inefficiently high according to the EPA's welfare accounting. Although the example is purely hypothetical, the critical analytical point is that there are no resources within the welfare-economic paradigm to avoid this conclusion. Just as it can appear to be welfare maximizing to cause future generations no longer to exist, society can appear to be "better off" with the mental capacity of its children deliberately reduced.

A similar confusion occurs with respect to the role of "victim avoidance behavior" in the economic analysis of environmental law.[27]

Occasionally, the harms of pollution and other environmentally impact-ful activities can be avoided either by reducing the activities themselves or by the adoption of avoidance behavior on the part of potential victims. In the mercury emissions rulemaking, for instance, a key issue raised by environmental justice advocates concerned the higher levels of contami-nated fish that are consumed by members of Great Lakes Native Ameri-can tribes and other subsistence fishers. Not surprisingly, exposure estimates reveal these individuals to face much greater health risks from mercury than predicted by the EPA's general population models. The question then arises: Should the tribes be made to abandon their tradi-tional practices if it would be monetarily cheaper for them to avoid being harmed than for society to avoid harming them? From the welfare-economic perspective, there is no a priori reason to prefer one or the other approach, given that everything depends on costs and benefits stripped of moral content. Indeed, the genius of Ronald Coase's storied article *The Problem of Social Cost* lies precisely in upsetting traditional ways of describing environmental problems through moralized terms of victim and polluter. Coase instead insists on seeing only reciprocal rela-tionships between parties who desire to use a resource in ways that cannot be jointly achieved.[28]

Nor did the EPA add much moral texture in its "environmental jus-tice" analysis of the final mercury rule: Rather than truly acknowledge that monetized costs and benefits cannot exhaust the normativity of policy making—rather than engage in a thorough inquiry into what it means, in moral and historical terms, to force Native American subsis-tence fishers to choose between their health and a cultural practice that is significantly constitutive of their identity—the EPA instead reported with apparent relief that the emissions limits it chose were not so strin-gent that tribes would be disproportionately *benefited* by the rule.[29] This passing moment in the EPA rulemaking betrayed the depth of ideologi-cal capture that cost-benefit analysis has wrought: Only an overextended form of welfare-economic reasoning—one that sees disembodied wel-fare impacts to the exclusion of the lives, communities, and histories that are visited by those impacts—would have identified unfair benefits to Native Americans as the chief environmental justice concern at issue for mercury. As Catherine O'Neill observes, the EPA's "approach rede-fines the relevant baseline: it ignores the current maldistribution of environmental benefits and burdens and erases the long history of efforts to colonize and assimilate Native peoples. If pursued seriously, such an inquiry would always disqualify efforts to make progress toward environmental justice by ameliorating disproportionate burdens."[30]

Consider the following case to illustrate how poorly justice claims fare when assimilated to the welfarist framework: In the context of antidiscrimination law, we have attempted with some success to force employers to hire, promote, and fire without regard to race and other protected statuses. In the absence of such legal protections, many individuals within disadvantaged groups historically have attempted to mask or "cover" their identity, in order to pass as members of the dominant group and receive equal, or at least better, treatment.[31] Today, covering continues most prominently with regard to sexuality, which is not yet protected by federal antidiscrimination law. To the welfare economist who is tasked with quantifying the costs and benefits of antidiscrimination law, the fact that individuals have less need to cover in order to receive fair employment treatment might be viewed as an identifiable and even quantifiable benefit of the law. Covering is psychologically costly, to say the least. But if the welfare analyst goes further to express concern that minority groups might be disproportionately "benefited" by antidiscrimination law—since members of the dominant group have fewer covering expenses to save—then the analyst will have lost sight of the moral ground for legislative action. If the analyst goes further still to suggest that covering might be a less expensive way to obtain equal employment treatment than antidiscrimination regulation, then—to echo Bernard Williams—the analyst will have had "one thought too many."[32] The analyst will have made the content of antidiscrimination law depend on its consequences, rather than having let the consequences flow from law's morally determined content.

Again, welfare economics does not have tools for recognizing when it has exceeded the bounds of its competency. As an indication of the degree of confusion in this area, some proponents of cost-benefit analysis attempt to deny that welfarism is a normative framework at all. They state, for instance, that their analysis of "risk equity will not be from the standpoint of moral criteria but rather social welfare maximization,"[33] the same standpoint that led the EPA to express concern over the possibility that its mercury rule might unfairly benefit Native American tribes. Similarly, proponents of cost-benefit analysis express a belief that "the moral debates" over discounting in the intergenerational policy context can be bracketed by "investigating people's actual preferences in this domain,"[34] an approach that implicitly deems the preferences of presently living individuals sovereign over all resources for all time, reducing even the lives of future individuals to the status of goods for present enjoyment. Like Robert Stavins's comments from the previous chapter, these analysts want to reassure observers that cost-benefit analysis consists only in a positive task of identifying and tabulating the welfare consequences of policy making. Any effort to actually define, construct, and implement a social welfare function, however, entails extraordinarily difficult normative judgments. The various dichotomies deployed throughout the economic paradigm—fairness versus welfare, equity versus efficiency, precaution versus maximization, sustainability versus optimality, and, ultimately, subjectivity versus rationality—only arise after the analyst has first defined the latter concepts using criteria that must include consideration of the former. No matter how rigorous its academic incarnation becomes, welfare economics contains no intellectual levees sufficient to prevent this spilling over of its practices into the normative.

*

These limitations suggest a need to reinvigorate environmental law, not merely in welfarist terms, but in terms of the ethical self-understanding of its authors.[35] As generally practiced, regulatory cost-benefit analysis offers a semblance of comprehensiveness and evenhandedness by arbitrarily normalizing the status quo distribution of rights and resources; in this way, environmental law's scope becomes limited to consideration of merely incremental changes to a sociolegal system that is otherwise unquestioned. In the welfare-economic sense, optimization comes to mean ever-more-refined tinkering with a given liberal market system, the unexamined rules of which help to form the set of constraints under which optimization occurs. The transformative potential of law is thus curtailed: Just as economics experienced a marginalist revolution in the nineteenth century, promoters of cost-benefit analysis hope to subject law to a marginalist revolution in the twenty-first. They face an insurmountable problem, however, in that the subject matter of law cannot be cabined in a way that would make cost-benefit analysis appear adequate to our needs. Law cannot self-negate in the way that economics has: Much of law is marginal or incremental in nature, to be sure, but unless potential moments of upheaval—of foundationalist reordering—are always latent within it, law ceases to be responsive to its subjects. Indeed, it ceases to be *law*, in that it no longer represents the self-expressed commitments of an integrated political community. To commit to law is to have the power to remake it.

[margin note: foundational reordering must be present in the law.]

Even the most philosophically nuanced defenders of cost-benefit analysis shy from this possibility. Matthew Adler and Eric Posner, for instance, avoid some of the most egregious overreachings of welfare-economic analysis, yet they still end up coming close to an endorsement of its hegemonic tendencies. They wisely deny that cost-benefit analysis is a framework with "bedrock moral significance," and instead see it only as a "decision procedure" that offers heuristic value as "a rough-and-ready proxy for overall well-being."[36] Likewise, they deny that overall well-being is morality's exclusive bedrock: On their account, "[morality] certainly includes overall welfare; but it may also include such factors as moral rights, the fair distribution of welfare, and even moral considerations wholly detached from welfare, such as intrinsic environmental values."[37] Nevertheless, despite these admirably restrained views, the authors still suggest that the outputs of welfare calculation should dominate law and policy, if not complete occupy them. This suggestion is most dramatically revealed by the authors' assertions that "overall welfare is the baseline goal of agencies"[38] and that any relevant nonwelfarist considerations are "more or less" already captured by the U.S. Constitution.[39]

[margin note: Constitution cabins welfare?]

If, in the cost-benefit state, important deontological and other non-welfarist considerations are to be found only in the Constitution, then we cannot understand much of the environmental law we already have. For instance, we cannot understand the NEPA's overriding command that we consider the rationality of our actions not just in terms of the consequences that they generate but in terms of the procedures that determine them. We cannot understand the health-based mandate of the CAA, which requires the EPA and the states to ensure an ample margin of protection from air pollution and which trucks no consideration of economic cost in doing so.[40] We cannot understand the ESA, the MMPA, the Wilderness Act, and numerous other legislative steps to cordon off certain life-forms and natural resources from the market's—and by extension welfarism's—demand that their worth constantly be demonstrated in order to avoid being transformed into just another factor of production. We cannot understand the Fishery Conservation and Management Act, which states plainly that although fishery conservation and management measures may "consider efficiency in the utilization of fishery resources . . . no such measure shall have economic allocation as its sole purpose."[41] Nor can we understand the *future* of environmental law, for, as argued throughout this book, we are increasingly being forced to examine the relationships that exist between our political community and various of environmental law's others. Populations such as future generations and nonhuman life-forms are both vulnerable to environmental harm and disabled from participating in the political communities that, in large part, determine their fate. They also are not included within the sphere of constitutional rights that Adler and Posner suggest exhausts the relevancy of deontology to law. Thus, according to the authors' approach, the alterity of these others would simply be accepted as part of the natural order, written into the axioms of welfare analysis, rather than treated as a constant call for ethical engagement.

Liberal constitutionalism must do better than this. The Enlightenment values of humanism, reason, and empiricism reach an apotheosis in welfarism, which offers the promise that our social contract can be finalized, written once and for all into a self-executing program of welfare maximization in which all value resides within individuals whose preferences—though sovereignly chosen—nevertheless can be objectively discerned through scientific observation. For welfare economics safely to serve the kinds of roles for which it has been promoted, the constitutional framework against which it operates must first be reformed to better reflect the lessons of ecology and the natural sciences and to better accommodate the need for continuing ethical conversation regarding the status of marginalized or unreachable interest holders. The promise of environmental law is that certain needs and interests of present and future generations, the global community, and other forms of life can be given foundational legal importance, in such a way that the ensuing costs and benefits that are observed by economists will reflect a *prior* determination by the political community to pursue environmental sustainability. Proponents of regulatory cost-benefit analysis instead try to force sustainability and intergenerational justice into a language of preexisting cost and benefit schedules that are simply "read" from an unsustainable and arbitrarily privileged status quo. Only if the market were *first* subjected to foundational constraints of ecological sustainability would its ensuing operations—its apparently "natural" cost and benefit schedules—begin to have the kind of empirical and normative standing that welfare economists presently afford them.[42]

Constitutional constraints placed upon resource use—such as a requirement that renewable resources only be used at their replacement rate and that nonrenewable resources only be exploited at the rate of development of renewable alternatives[43]—would provide much of the foundationalist reordering necessary for welfare economics to become less misleading in the environmental law and policy context. Such constraints need not be thought of as radical or otherwise discontinuous with the liberal constitutional tradition. Instead, they could be seen as serving the largely anthropocentric and welfarist purpose of ensuring an ecologically sustainable path of development, one that no longer depends on optimistic suppositions that "enough, and as good" resources will always remain available for future generations, that all resources are substitutable even if particular scarcities do arise, or that societies inevitably "grow" their way to a "preference" for environmental conservation. Understood in this way, the seemingly dramatic alteration worked by sustainability constraints would not actually constitute a breach with the Enlightenment traditions of humanism, reason, and empiricism. Instead, it would simply reflect a different preanalytic orientation toward science (that is, more precautionary) and an expanded horizon of moral and political concern (that is, one that includes future individuals).

An alternative conceptualization of sustainability constraints—one that refuses to see the constraints as serving the needs of *particular* rights-bearing individuals, even future individuals—would pose a more direct challenge to the Enlightenment tradition. From this perspective, sustainability constraints would be seen as absolute limitations on public and private power: The limit they impose would concern the exploitation of resources that sustain life and that are held in trust for all life.[44] As has been noted, many of environmental law's subjects are not politically represented in the usual liberal fashion. Instead, they are glimpsed through a conjunction of science and ethics that places both knowledge schemes under tremendous strain. Indeed, environmentalism often appears unscientific and illiberal precisely because the true subjects of its scientific and ethical inquiry are unrecognized by skeptical observers.[45] In light of this dilemma, commentators frequently contend that environmental law's ultimate ambitions—those that go beyond protection of the environment for anthropocentric reasons—are likely to fail, given that they "do not draw upon the philosophical, religious, and jurisprudential bases of the constitution, all of which are rooted in the enhancement of human dignity."[46]

This conclusion is premature. Much like the notions of territory and sovereignty that have been used to stabilize international ordering, the concept of human dignity that has been used to organize constitutional frameworks is called into question by the tasks of environmental law. Uncomfortable though it may be to admit, we do not yet know whether our intellectual levees will hold and, if they do not, in which direction they will yield. For instance, how will juxtaposed categories of society and nature, persons and things, autonomy and allonomy be understood once their hybridity is laid bare by genetic engineering technologies? Will concepts of freedom and choice have their familiar significations for subjects who see their maker not in a transcendent deity or an intergenerational secular compact, but in a life sciences corporation? Can reason and empiricism still provide liberalism's neutral public vocabulary when they confront us with the charge that our chosen lifestyles are affecting the environment on a geological—indeed a biblical—scale? Will the demarcation between public and private spheres so essential to liberalism still hold when activities in the latter are seen to be binding the entire planet into a shared and possibly dismal climatic fate?

Even in its more theoretically bold formulation, the attempt by environmental law to include others should not be seen as a breach of the liberal tradition, but instead as a strengthening of it. In an ideal discourse community that extends across boundaries of space, time, and speciation, environmental law's others would themselves be present, not merely represented. Their faces would be visible, and their needs unmistakable. In the absence of such an idealized situation, we must pursue practical methods of expanding environmental impact assessment and natural resource planning in order to begin a process of recognition.

Would-be interest holders might be better recognized as such, and might eventually be invited into the political community as members, if we were constitutionally required to assess our biological inventory and natural resources, to monitor the deleterious impacts of our activities on life forms outside our territorial borders, to imagine our future population and the state of their environment, and to consider thoughtfully our domesticated and engineered life forms and the quality of their existence. It is not enough to simply assert that "nations lack faces"[47] as an argument that we are incapable of regarding international relations with moral sentiment. Instead, we must try to *give nations faces,* just as we must seek to better recognize and respect non-human life forms.[48] Likewise, if the Constitution is, as Bruce Ackerman describes, a conversation between generations,[49] then we must seek to better hear the diachronic expressions of future generations.

This understanding of environmental constitutionalism—partly as a structural endeavor and partly as a device of recognition—is preferable to the focus on individual human rights that has dominated discussions of environmental constitutionalism. At present, courts and commentators view environmental constitutionalism as a precarious effort because they do not see it as being of the same order as established features of constitutional law. For instance, as Thompson notes, U.S. state courts have tended to shy away from adjudicating environmental constitutional provisions because, unlike rights such as freedom of speech and due process, courts do not see environmental provisions as being "cornerstones for an effective representative democracy."[50] Viewed through the appropriate lens, however, environmental provisions would also be recognized as prerequisites to efficacious government: They would reflect the fact that no liberal political community should ever view itself as completed, that the community instead should always question whether its vision of harmonious self-ordering could be made to be more inclusive. To be sure, identifying the proper contours of a fair framework of social cooperation is a grand challenge even for a single generation of acknowledged co-citizens. Nevertheless, any comprehensive political theory—and any actually existing political community—must also seek to establish a fair framework for cooperating with members of *other* communities, including the possibility of embracing them wholly as equal subjects.[51]

[handwritten margin note: Env. constit. preferable to focus on human rights...]

[handwritten margin note: prereqs. to efficacious government]

Because of its relentless attention to interrelation, environmental law can be of great use in this project: Environmental law raises issues of dependency and membership that haunt the liberal vision in just the right way. Too often in the liberal tradition environmental law's others are simply taken to reside outside the conditions of justice; as a result, we see only meager substantive provision for their needs, even in powerful manifestations of liberal theory, let alone in lived experience. Likewise, these entities express no willingness-to-pay of their own within economic cost-benefit analysis, but instead must always be *paid for* by a life that has already somehow been granted standing in the cost-benefit community. As argued throughout this book, these exclusions tend to rest on historical accident and practical concession rather than on persuasive moral reasoning. Indeed, much like the early welfare economists who regarded discounting as a regrettable convenience, those who conclude that the conditions of socioeconomic justice do not prevail between states tend do so sheepishly, acknowledging that the cosmopolitan alternative has obvious normative appeal. In both cases, theorists provide an ethically compromised final response to a problem that is neither finally resolvable nor ethically negotiable. We cannot, as yet or ever, determine the substantive content of global justice, just as we cannot predict the needs and preferences of future generations or adequately describe the awesomeness of species extinction. But neither can we ignore these calls.

Accordingly, liberal constitutionalism should not seek final answers to questions of recognition and membership. Instead, it should seek to reinforce collective self-consciousness regarding the need to confront those questions and to remain always dissatisfied with their instant resolution. When individuals bind together to form a political community, they create more than a set of institutions and procedures to maximize individual interests. They also create a collective agent that holds an inevitable relationality with *other* collective agents.[52] What justice requires between those collectivities cannot simply be fixed according to a global welfare calculus, reduced to an algorithm of recognition, or excluded as outside the scope of justice theory. Just as the individual cannot sublimate agency either to a consequentialist-utilitarian program or to a set of inviolable deontological maxims, but instead must always and in every context imperfectly *decide*, so too must collective agents always bear a

burden of unavoidable, yet unsatisfiable responsibility. Liberal constitutionalism should therefore view itself always as a work in progress,
asymptotically striving toward an unattainable but undeniable goal of
universal recognition and respect.

Something of the requisite restlessness is captured in an oft-quoted
Samuel Beckett passage, particularly its final two words: "Ever tried.
Ever failed. No matter. Try again. Fail again. Fail better."[53] Beckett's admonition to "fail better" powerfully reminds us that limitations of knowledge and power do not erase the responsibility to strive. Indeed, they are
part of the very space in which agency and responsibility are born. As
Jacques Derrida wrote—in words that merit constant consultation: "[A]
decision can only come into being in a space that exceeds the calculable
program that would destroy all responsibility. . . . [T]here can be no
moral or political responsibility without this trial and this passage by
way of the undecidable."[54] Yet the space or passage separating choice
and consequence is precisely what cost-benefit analysis seeks to close.
Policies, we are told, do not result from the judgment of a responsible
subject but instead "are hostage to what the facts turn out to show in particular domains." No air reaches this closed passage. Try as we might to
inhabit the calculable program of welfare economics, the necessity of
deciding amidst conditions of undecidability remains and in fact under-
writes the very meaning of environmental law. Undecidability explains
both the failure of the welfare-economic attempt to reduce environmental lawmaking to objectivist technique and the wisdom of the precautionary principle's implicit reminder that we are a political community
whose judgments must include a leap—the same nonprogrammable
leap that motivates a commitment to execute our judgments with care
and with honor. They are, after all, *our* judgments.

*

This chapter would not be incremental and pragmatic in pitch if it ended
here. Whatever the theoretical power of the case for environmental constitutionalism, attention also must be paid to the worldly context within
which such a transformation must occur. Again, there is cause for doubt:
As part of a skeptical assessment of the prospects for an environmental
amendment to the U.S. Constitution, J. B. Ruhl offers the sobering reminder that more than ten thousand constitutional amendments have
been proposed in the nation's brief history, yet only a small handful have

survived Article V's daunting procedure for amendment.[55] To assess the likelihood of success for an environmental amendment, Ruhl first maps constitutional provisions on a biaxial matrix based on their function and target. Ruhl's function axis classifies a constitutional provision according to its practical aim: "(1) altering the operational rules of government; (2) prohibiting specified government action; (3) creating or affirming rights; or (4) expressing aspirational goals." His target axis evaluates what relations are governed by the provision: "(1) intra- and intergovernmental relations; (2) relations between government and citizen; or (3) relations between citizens."[56] As Ruhl notes, the further along these axes a provision aims, the less likely it seems to be capable of permeating constitutional culture, at least as judged by past experience. Most notably, provisions that offer aspirational goals or that target relations between citizens have been notoriously unsuccessful. He concludes that an individual right to a clean and healthy environment scores poorly on this constitutional matrix, given that its substantive content necessarily appears aspirational in advance of social consensus on appropriate levels of environmental protection and that, as a practical matter, its target must include relations that are understood to lie solely between private persons.

As argued throughout this chapter, however, environmental constitutionalism can be reframed to avoid the perception that it rests so uneasily within the liberal constitutional tradition. At its core, environmental constitutionalism seeks to accomplish two improvements to liberal thinking, neither of which needs to be seen as radical or discontinuous. First, it aims to substitute a conservative assumption of finitude and nonsubstitutability for the hope that "enough, and as good" resources exist to justify largely unregulated privatization and use of property. This ecological assumption need not be considered antimarket or value-coercive: As Herman Daly has emphasized, sustainability constraints on the use of renewable and nonrenewable resources, including the pollution-absorptive capacity of the atmosphere and other natural sinks, can be implemented through tradable permit schemes in a manner that continues to promote allocative efficiency through decentralized private decision making.[57] Optimization simply occurs under an improved set of constraints. Second, environmental constitutionalism aims to force liberalism to become more self-conscious of its membership decisions. Historically the push beyond anthropocentricism has been the most noted membership challenge issued by environmentalism; indeed, it is this ecocentric feature that has caused many observers to be skeptical of the movement's ability to effect constitutional change. But environmentalism, with its ceaseless attention to spatial, temporal, and biological interconnection, simply poses the kinds of membership questions that liberalism already faces and that it no longer can tenably avoid. Environmental constitutionalism would not force us to abandon cherished notions of human dignity; it simply would force us to abandon complacency about whether we have drawn the circle of dignity as widely as possible. In short, rather than a vague or hortatory extension of individual rights into the realm of affirmative entitlements, environmental constitutionalism instead can be seen to rest comfortably beside more firmly established structural and representative aspects of constitutionalism.

Nor must environmental constitutionalism be understood to depend on the cumbersome constitutional amendment process or on heroic acts of judicial interpretation. As Bruce Ackerman has argued through a series of compelling works, constitutionalism must be envisioned as living beyond the four corners of the written text and the four walls of the Supreme Court.[58] In some cases, foundational legal change occurs through more diverse and diffuse mediums, including through movement parties, media politics, dramatic shared social experiences, highly symbolized presidential elections, and the passage of landmark statutes such as the Social Security Act or the Civil Rights Act. Although difficult to model as a causal system, this broader array of constitutional determinants can, under certain circumstances, come together to enable the proposal, acceptance, and consolidation of a constitutional change just as powerfully as the more conventionally studied mechanisms of judicial interpretation and formal amendment. Indeed, as Ackerman shows, even the formal court opinions and written amendments that traditionally have dominated our constitutional canon can be better understood within this larger institutional and cultural context. For instance, the full meaning of the nation's revised understanding of equal protection and state responsibility that emerged in the second half of the twentieth century is surely not to be found in the Supreme Court's limp acceptance of the constitutionality of the Civil Rights Act on interstate commerce grounds.[59]

Might we profitably understand 1970s-era federal environmental statutes from this perspective? After all, the environmental movement modeled many of its litigation, media, and advocacy strategies on the civil rights movement; broad acceptance of environmental goals and values within the culture coincided with similarly progressive changes in attitudes toward race and equality; environmental legislation enjoyed extraordinary levels of bipartisan support in Congress; and Richard Nixon's creation of the EPA and his general acceptance of environmental policies seemed to seal the foundational legitimacy of the movement, just as his refusal to oppose the Civil Rights Act helped to signal the ultimate constitutional success of the Second Reconstruction.[60] Also like federal civil rights legislation, many of the landmark federal environmental laws seem on their face to aspire to something more dramatic than business-as-usual lawmaking.[61] For instance, unless one is already aware of its subsequent judicial curtailment, the NEPA statute can only be read as a deliberate effort to revolutionize the approach of agencies and courts to the implementation of *all law*—an aspirant to superstatute status if ever there was one. Indeed, though scarcely noted today, the statute contains a provision that could easily have been read to confer a protoconstitutional right to environmental protection.[62] Equally ambitious from the expanded constitutionalist perspective are the ESA, which prohibits not only government actors from contributing to species loss, but private actors as well; the CWA, which flatly prohibits the degradation of high quality waters, irrespective of the opportunity costs of doing so; and the CAA, which vests far-reaching regulatory power in a federal agency, the full breadth of which only became apparent after the EPA began considering how it might implement a greenhouse gas emissions control program following *Massachusetts v. EPA*.[63]

It is certainly true, as Ackerman cautions, that "[m]ost statutes and executive decrees simply don't proceed from the sustained deliberation typical of our great acts of popular sovereignty."[64] Thus, most laws and regulations do not merit canonization in our expanded constitutional understanding. Nevertheless, unlike garden variety statutes and orders, much of federal environmental lawmaking *was* on a course for constitutional significance, until a rash of restrictive court interpretations and a changing political environment resulted in a downgrading of its potential power and meaning. Matters were not helped when the oil shock and the economic crisis of the late 1970s caused environmentalism to become a much more polarizing and uncertain national value. Had the United States instead remained popularly committed to the reformist vision found within its environmental statutes, then one suspects that the EPA and the courts would not have interpreted those statutes as narrowly as they have. Likewise, had the statutes been fully pursued and enforced, then they would not appear so idealistic and naïve to us today, for the technologies, knowledges, and values that we would be using to assess them would be quite different. Nor would the statutes be easily dismissed as inefficient or irrational according to cost-benefit analysis, since we would be unable to ignore their ambition to alter the underlying systems that give rise to cost-benefit values, rather than to simply respond to those values. Environmental law would neither operate, nor be evaluated, at the margins. It would instead live within our constitutional conversation.

Recognizing that the cultural, political, and legal momentum necessary to consolidate this environmental constitutionalist effort has waned, this book closes with a simple, but hopefully catalytic legislative proposal: the Environmental Possibilities Act. The act seeks to capitalize on renewed popular concern for the environment. After nearly three decades in which scientific distortion and misleading economic analysis obscured the significance of anthropogenic greenhouse gas emissions, the problem of climate change now draws wide attention, even among groups not generally disposed to favor environmental policies. The dominant policy framework within which climate change is being addressed, however, fails both to confront the fullness of the climate change problem and to consider the range of imaginative and dynamic responses we might take to address it. The framework applies static and partial efficiency analysis to a problem that is unequivocally dynamic and general in scope. Indeed, were it not for successful efforts by the OMB, conservative think tanks, and academic commentators to promote cost-benefit analysis more broadly over the past three decades, one would wonder how we ever came to think that the welfare-economic framework was appropriately suited to the climate change problem. Climate change is, in fact, a poster child for the limitations of cost-benefit analysis.

Accordingly, a central aim of the Environmental Possibilities Act is to resituate the government's relationship to cost-benefit analysis, so that environmental problems such as climate change need not be addressed from within a conceptual straitjacket. The act aims to institutionalize the forgotten wisdoms of the precautionary approach to environmental law, including most importantly the virtues of humility and self-awareness that lie at its core. More speculatively, the act aims to provide a focal point for popular remobilization of our one-time environmental constitutionalist agenda, believing that the transformations we had begun remain the transformations we need.

*

Environmental law concerns most basically life and the conditions that support it. Yet despite its deep connections with romantic and transcendentalist traditions, and despite its modern beginnings as a transformative, even quasi-constitutional field, environmental law in recent years has shifted toward a technical orientation in which life is treated instrumentally, as a resource to be deployed in service of overall well-being. Rather than protect life proper, environmental law has reduced its focus to what Giorgio Agamben calls "bare life," those aspects of life that can be apprehended from a scientific, instrumentalist, and bureaucratic perspective. Not only wetlands, whales, and oil wells, but human beings are treated as resources to be managed. All lives and all values become objects of study and manipulation. From this perspective, "natural life itself and its well-being seem to appear as humanity's last historical task."[65] The careful distinction that Amartya Sen insists upon—between viewing humans as "agents whose freedoms matter" rather than as "patients who are no more than their living standards"[66]—vanishes, as does any hope of expanding the class of freedom-seeking agents beyond the instant, acknowledged group of individual humans.

Repeatedly, we are told that this instrumentalization of life is necessary for our own good and that we will be "better off" for having sublimated our agency to managerial expertise. But, as Agamben writes, it is far from "clear whether the well-being of a life that can no longer be recognized as either human or animal can be felt as fulfilling."[67] As a seemingly frivolous example, consider the fact that consumers in the United States spend more per year on their household pets than the gross domestic product of most countries.[68] Consider also the fact that pharmaceutical companies in the United States have successfully opened a market for behavioral modification aids for those pets, including formulations of Prozac and other psychotropic drugs for cats and dogs believed by their owners to be too energetic.[69] Are these signs that the prejudicial distinctions between "man" and "cat" are eroding in response to the ethical call of the other? Or are they signs that the instrumentalization of all life is one step closer to completion? If so, will it be fulfilling for them? For us?

Coming genetic engineering technologies will reveal the folly—and danger—of the approach to life as bare life. As argued in chapter 7, our dominant moral and political traditions lack adequate means to guide the use of such technologies, since they depend on the very kinds of categorical distinctions that will be brought into question by our powers of design and control. How will we respond when the foundational tools of those traditions are set against each other? Will our confidence in reason and empiricism, articulated through science and technology, lead us further down the path of life's instrumentalization, or will our commitment to the human individual reignite upon seeing too much of life reduced to the status of a patient? Will the vacant gaze of our medicated cat appear to us as a mirror? What will be our test for moral and political recognition when we no longer have recourse to any notion of a naturalistic order but instead must measure directly the ethical ambition within us? These tensions have always been with us—we have, in that sense, never been modern[70]—but the genetic age will bring them to the fore with undeniable clarity.

Clear also will be the point of precaution. This book began by arguing that the precautionary principle and associated regulatory approaches from the dawn of modern environmental law embody a variety of wisdoms that have become endangered amidst the dominance of the welfare-economic program. The most basic of wisdoms remains the implicit reminder that how a society enacts its causal capacity—in geopolitical, intertemporal, and interspecial terms—will determine the content of its identity. Understandably, such self-consciously collective political conceptions tend to be regarded with distrust. Environmental law becomes guilty by association: In 2007, for instance, with the world's attention focused on the challenge of climate change, the president of the Czech Republic, Václav Klaus, claimed to "see the biggest threat to freedom, democracy, the market economy and prosperity now in ambitious environmentalism, not in communism."[71] The mistake of commentators like Klaus, however, is to believe that polities must choose exclusively between liberal individualist and collectivist self-conceptions. The liberal individualist framework might be most needed to govern relations internal to a polity, where minority groups and other marginalized denizens might easily be ignored or abused by a majoritarian political culture that becomes too accustomed to declaring what "We the people" embody in thick, personified terms. But to extend this framework outward to govern also the polity's relations with missing interest holders, like foreign nations and future generations, is a mistake: The same individualism that works to protect preexisting members of a community from being dismissed or maligned as other *works an othering* when applied to those who are not already self-present members of the political community. "We the people" must therefore admit of a more collective self-consciousness when structuring relations with environmental others, or else we will fail to appreciate the historical scale in which those relations unfold.

Perhaps sharing some of Klaus's views, promoters of welfare economics defend their framework on the ground that it promises to reduce discretion, judgment, and other trappings of robust collective agency. They hope to eliminate such agency because they believe that the view of regulation from nowhere—the objective view that attaches no special significance to the identity of the political community that is engaged in policy making—best avoids oppressive, paternalistic, corrupt, or otherwise misguided government action. Whatever its success at achieving those aims, the economic framework has taken us too far in the direction of nowhere, causing us to lose sight of ourselves and our challenges by promising that the normativity of environmental law can be determined through empirical assessment of welfare consequences alone. Because the precautionary principle "resist[s] the voracity of the objective appetite,"[72] it cannot be dismissed with the haste its critics urge. As argued throughout this book, the precautionary principle is of a piece with the long-recited intonation—"First, do no harm"—which works to remind the physician to be aware of her special role and causal capacity, just as the precautionary principle serves to remind the political community that it is a distinctive actor within history. The language of such demands is not the literal one of a program of action, but the more ancient one of a reminder of responsibility that cannot be shed, ethics that cannot be systematized. Problems at the frontiers of justice demand more in the way of humility, striving, and unflagging self-awareness than they do in scientistic or rationalist rigor. They require awe at the sheer power of being within history, of being able to influence history by connecting thought to action in a world that is already otherwise connected and that, accordingly, will react in ways we cannot foresee and often will not desire. Taken to an extreme, such awe would lead us, like Jainists, to gingerly step behind broomed surfaces, lest we make moves that cause death even to the lives that lie beneath the reach of our vision. But the basic wisdom—the essential wisdom—is that causing death must always be done with solemnity and with the hope of greater precaution tomorrow.

We are facing a global environmental crisis, but its root causes are epistemological, ethical, and political in nature: epistemological in the sense that, since the Enlightenment, our institutionalized search for knowledge has been largely reductionist and empirical in orientation, yet now we are confronted by challenges whose full magnitude is only perceptible—let alone resolvable—through integrative and imaginative thinking; ethical in the sense that our dominant models of human behavior at present fixate on interest maximization in an effort merely to describe and predict, yet our models and their expressions have begun endogenously to limit our understanding of ourselves and our relations; and political in the sense that, also since the Enlightenment, we have inhabited conceptions of organization that are primarily individualistic and insular in form, yet now we require conceptions that admit smoothly into their purview statistical victims, foreign citizens, future generations, nonhuman life-forms, and other seemingly absent interest holders. The global environmental crisis is therefore in no small measure a crisis of ideas.

Resolving our intellectual crisis and ensuring environmental sustainability will require both humility regarding our powers of prediction and control, and courage regarding our ability to engage in a form of public decision making that conceives of political communities as actors with their own agency, responsibility, and history. The risks of oppression raised by such a vision are well known and justly feared. But we have overlearned the lesson of the twentieth century. It is and will remain vital to guard against the impoverishing and oppressive possibilities that lurk within governance mechanisms that are self-consciously collective, yet the opposite dream—the dream that a fully just, efficient, and sustainable planet can emerge more or less spontaneously from the liberal order—is a nightmare of its own. It encourages cornucopian fallacies that raise the stakes of our onetime global experiment through an ever-expanding scale of human activity, even as they deny the need to address our environmental distress through more direct and courageous measures.

Entranced by our powers of reason and observation, we have sought only the indubitable and forgotten the undeniable. The undeniable, if unfulfillable, promise of the Enlightenment is simply this: Let every life, let every death, speak for itself.

APPENDIX: THE ENVIRONMENTAL POSSIBILITIES ACT

AN ACT

To reaffirm and strengthen the principles espoused in the National Environmental Policy Act of 1969; to acknowledge and embrace the central meaning of the precautionary principle, which is that the nation stands in a relationship of responsibility to present and future generations, to members of foreign nations, and to other living beings; and to declare a national commitment to hold open possibilities for the environment and for the lives that it supports.

Be it enacted by the Senate and House of Representatives of the United States of America in Congress assembled.

SECTION 1. SHORT TITLE

This Act may be cited as the "Environmental Possibilities Act."

SECTION 2. FINDINGS

The Congress finds that:

(a) The nation's environmental, health, and safety laws and regulations have led to improvements in the environment and have significantly reduced human health risk; however, the nation and the world continue to face serious challenges in ensuring a mode and level of development that is prosperous, inclusive, and environmentally sustainable.

(b) The increasing technical sophistication of environmental policy making has discouraged public engagement with the inevitable gaps, uncertainties, and ambiguities of scientific and economic assessment, as well as the moral and political values that are implicated by such assessment.

(c) The use of economic cost-benefit analysis in particular has promoted a narrow understanding of environmental, health, and safety law, one that pursues the maximization of well-being for certain individuals using certain criteria of value, rather than undertaking a broader effort to hold open possibilities for flourishing, not only for presently living citizens but also for members of future generations and foreign nations, for nonhuman life-forms, and for the environment itself.

(d) While it is inevitable that policy choices involve tradeoffs between competing considerations, the following matters are not inevitable: that the nation become complacent about the manner in which tradeoffs are resolved; that lives lost or harms caused be accepted rather than regretted; that the institutions, technologies, and processes that structure any momentary depiction of inevitable tradeoffs themselves lie beyond reform; or that there be no advantage in seeking to fail better, rather than to optimize, when society aims to protect life and the environment.

SECTION 3. PRINCIPLES FOR ECONOMIC COST-BENEFIT ANALYSIS

(a) IN GENERAL—The head of each federal agency shall apply the principles set forth in subsection (b) in order to ensure that any economic cost-benefit analysis it uses or conducts, including any analysis required under any federal law or executive order, promotes effective and transparent decision making, free of undue attachment to any particular and possibly nonrepresentative set of moral and political values.

(b) PRINCIPLES—The principles to be applied are as follows:

(1) PRELIMINARY JUDGMENT—Before using or conducting an economic cost-benefit analysis, the agency shall first ascertain whether the statute being implemented is appropriate for the use of economic cost-benefit analysis at all. Relevant factors shall include: whether the Congress considered and rejected economic cost-benefit analysis when adopting the statute; whether the statute requires policy approaches, such as health-based, technology-based, or feasibility-based standards, that are analytically distinct from cost-benefit analysis; whether the statute evinces a congressional intent to alter the underlying market structure that gives rise to conventional cost-benefit valuations, rather than to have the statute's implementation be determined by those valuations; whether existing scientific and empirical understandings of the problem to be addressed are adequate to plausibly support the cost-benefit methodology; and whether the agency's attempt to identify costs and benefits for policy analysis is likely to be frustrated by endogenous effects of the policy itself.

(2) WELFARE CRITERIA—If, when engaging in an economic cost-benefit analysis, the agency chooses to calculate welfare consequences using willingness-to-pay measures of value, the agency shall also present calculations using alternative measures of value, such as willingness-to-accept (when likely to diverge significantly from willingness-to-pay), equity-weighted welfare functions, or objective measures, such as those underlying the United Nations Human Development Index. In addition to

presenting such alternative calculations, the agency shall also provide a clear conceptual explanation of each value criterion used.

(3) INTERGENERATIONAL EFFECTS—For any economic cost-benefit analysis in which significant intergenerational effects are at issue, the agency shall refrain from discounting future costs and benefits to a present value based on the pure rate-of-time preference or any other factor relating solely to temporality. If the agency chooses to discount future costs and benefits to reflect opportunity costs, such as through the private cost of capital or a social discount rate, the agency shall present multiple calculations using alternative discount factors and shall clearly explain the conceptual basis for each factor. When the monetized value of future generations' lives or health is being discounted to a present value, the agency shall state so clearly and shall also present nondiscounted calculations. When significant natural resources or ecosystem services are at issue, the agency shall present alternative calculations in which such resources or services are assumed to belong at least in part to future generations, and in which relevant cost and benefit values are derived through modeling of markets that include such intergenerational ownership constraints.

(4) EXTRATERRITORIAL EFFECTS—For any economic cost-benefit analysis in which significant extraterritorial effects are at issue, the agency shall present multiple calculations using alternative approaches to the questions of whether and how to value such extraterritorial effects, including an alternative in which extraterritorial effects are valued no differently than domestic effects are. The agency shall provide a clear conceptual explanation of each alternative approach used.

(5) NONQUANTIFIED BENEFITS—For any economic cost-benefit analysis in which nonquantified environmental, health, or safety benefits are likely to be significant in relation to the overall set of consequences being evaluated, the agency shall refrain from specifying a quantified net cost-benefit outcome in its analysis.

SECTION 4. DECISION CRITERIA

(a) IN GENERAL—No federal agency shall promulgate a final rule unless the agency certifies, in a statement published in the *Federal Register* concurrently with the issuance of the rule, that any economic cost-benefit analyses used or conducted by the agency in connection with the rule are in compliance with the principles of section 3.

(b) COMMITMENT TO FAIL BETTER—In the case of any rule, the primary object of which is environmental, health, and safety protection, the statement referred to in subsection (a) shall also include the agency's certification that the final rule being promulgated contains significant substantive or procedural requirements in furtherance of environmental, health, and safety protection that have not been derived through comparison of incremental costs and benefits alone, irrespective of the method for calculating such costs and benefits.

(C) EFFECT OF DECISION CRITERIA—Notwithstanding any other provision of federal law, the decision criteria of subsections (a) and (b) shall supplement and, to the extent that there is a conflict, supersede the decision criteria for rulemaking otherwise applicable under the statute pursuant to which the rule is being promulgated.

SECTION 5. JUDICIAL REVIEW

Any interested person shall have the right to commence a civil suit to enforce compliance by a federal agency with the requirements of this Environmental Possibilities Act. The United States district courts shall have jurisdiction to hear such action without regard to the amount in controversy or the citizenship of the parties. In the event that a party successfully demonstrates noncompliance by an agency with the requirements of this act, the court shall enforce the provisions of this act as appropriate through injunctive relief and shall award attorney fees and other costs of litigation to the complaining party.

An "interested person" for purposes of this section means any natural person or any living entity desiring to be recognized as a natural person; in case of doubt regarding the desire of such living entity, the court shall appoint a recognized natural person as guardian ad litem and permit the suit to proceed.

NOTES

INTRODUCTION

1. The literature criticizing economic approaches to environmental law and welfare economics more generally is vast. For a sampling of important contributions, *see* FRANK ACKERMAN & LISA HEINZERLING, PRICELESS: ON KNOWING THE PRICE OF EVERYTHING AND THE VALUE OF NOTHING (2004); SIDNEY A. SHAPIRO & ROBERT L. GLICKSMAN, RISK REGULATION AT RISK: RESTORING A PRAGMATIC APPROACH (2003); MARK SAGOFF, THE ECONOMY OF THE EARTH: PHILOSOPHY, LAW, AND THE ENVIRONMENT (1988); Amy Sinden, *Cass Sunstein's Cost-Benefit Lite: Economics for Liberals,* 29 COLUMBIA J. ENVTL. L. 191 (2004); Thomas O. McGarity, *A Cost-Benefit State,* 50 ADMIN. L. REV. 7 (1998); ELIZABETH ANDERSON, VALUE IN ETHICS AND ECONOMICS (1993); Duncan Kennedy, *Cost-Benefit Analysis of Entitlement Problems: A Critique,* 33 STAN. L. REV. 387 (1981); Steven Kelman, *Cost-Benefit Analysis: An Ethical Critique,* REGULATION, Jan./Feb. 1981, at 33; Mark Kelman, *Choice and Utility,* 1979 WIS. L. REV. 769 (1979); Mark Kelman, *Consumption Theory, Production Theory, and Ideology in the Coase Theorem,* 52 S. CAL. L. REV. 669 (1979); Laurence H. Tribe, *Policy Science: Analysis or Ideology?* 2 PHIL. & PUB. AFF. 66 (1972). For recent and important attempts to recast economic analysis in less divisive terms, *see* RICHARD L. REVESZ & MICHAEL A. LIVERMORE, RETAKING RATIONALITY: HOW COST-BENEFIT ANALYSIS CAN BETTER PROTECT THE ENVIRONMENT AND OUR HEALTH (2008); and MATTHEW D. ADLER & ERIC A. POSNER, NEW FOUNDATIONS OF COST-BENEFIT ANALYSIS (2006).

2. *See* RICHARD LAZARUS, THE MAKING OF ENVIRONMENTAL LAW (2004).

3. *See* Thomas O. McGarity, *The Goals of Environmental Legislation,* 31 B. C. ENVTL. AFF. L. REV. 529, 538–45 (2004) (reviewing examples).

4. *See* David M. Driesen, *Distributing the Costs of Environmental, Health, and Safety Protection: The Feasibility Principle, Cost-Benefit Analysis, and Regulatory Reform*, 32 B. C. ENVTL. AFF. L. REV. 1 (2005).

5. *See, e.g.*, U.S. ENVIRONMENTAL PROTECTION AGENCY, THE BENEFITS AND COSTS OF THE CLEAN AIR ACT, 1970 TO 1990, at ES-8 (1997), *available at* http://www.epa.gov/air/sect812/812exec2.pdf (finding compliance costs of more than $500 billion but benefits in excess of $22 trillion). *See also* Shi-Ling Hsu, *Fairness Versus Efficiency in Environmental Law*, 31 ECOLOGY L.Q. 303, 342 (2004) (gathering studies).

6. For two of the most powerful academic statements of the reform agenda from this period, *see* Bruce A. Ackerman & Richard B. Stewart, *Reforming Environmental Law*, 37 STAN. L. REV. 1333 (1985), and STEPHEN BREYER, BREAKING THE VICIOUS CIRCLE: TOWARD EFFECTIVE RISK REGULATION (1993).

7. This common refrain actually mischaracterizes technology-based regulations in a significant way. In almost all cases, technology-based regulations identify a single technology only for purposes of identifying the standard of performance that regulated firms must achieve, leaving those firms free to choose the most cost-effective way of achieving it.

8. *See* Sam Peltzman, *The Regulation of Auto Safety*, in AUTO SAFETY REGULATION: THE CURE OR THE PROBLEM? 2 (Henry G. Manne & Roger LeRoy Miller eds., 1976); W. Kip Viscusi, *The Lulling Effect: The Impact of Child-Resistant Packaging on Aspirin and Analgesic Ingestion*, 74 AM. ECON. REV. 324 (1984). Elsewhere, Viscusi refers to the "lulling effect" as the more general phenomenon in which the introduction of government safety regulations "produce[s] misperceptions that lead consumers to reduce their safety precautions because they overestimate the product's safety." W. KIP VISCUSI, FATAL TRADEOFFS: PUBLIC AND PRIVATE RESPONSIBILITIES FOR RISK 234–42 (1992).

9. For key contributions to the risk-risk literature, *see* Jonathan B. Wiener, *Precaution in a Multi-risk World*, in HUMAN AND ECOLOGICAL RISK ASSESSMENT: THEORY AND PRACTICE 1509 (Dennis D. Paustenbach ed., 2002); Jonathan Baert Weiner, *Managing the Iatrogenic Risks of Risk Management*, 9 RISK: HEALTH, SAFETY & ENV'T 39, 40 (1998); John D. Graham & Jonathan Baert Wiener, *Confronting Risk Tradeoffs*, in RISK V. RISK: TRADEOFFS IN PROTECTING HEALTH AND THE ENVIRONMENT 1 (John D. Graham & Jonathan Baert Wiener eds., 1995); W. Kip Viscusi, *Risk-Risk Analysis*, 8 J. RISK & UNCERTAINTY 5 (1994); AARON WILDAVSKY, SEARCHING FOR SAFETY 212 (1988); Chauncey Starr & Christopher Whipple, *Risks of Risk Decisions*, 208 SCI. 1114 (1980).

For contributions to the health-health literature, *see* W. Kip Viscusi, *Regulating the Regulators*, 63 U. CHI. L. REV. 1423, 1452–53 (1996); Randall Lutter & John F. Morrall III, *Health-Health Analysis: A New Way to Evaluate Health and Safety Regulation*, 8 J. RISK & UNCERTAINTY 44 (1994); Randall Lutter, John F. Morrall III & W. Kip Viscusi, *The Cost-Per-Life-Saved Cutoff for Safety-Enhancing Regulations*, 37 ECON. INQ. 599 (1999); Frank B. Cross, *When Envi-*

ronmental Regulations Kill: The Role of Health/Health Analysis, 22 ECOLOGY L.Q. 729 (1995); Ralph L. Keeney, *Mortality Risks Induced by the Costs of Regulations,* 8 J. RISK & UNCERTAINTY 95 (1994); W. Kip Viscusi & Richard J. Zeckhauser, *The Fatality and Injury Costs of Expenditures,* 8 J. RISK & UNCERTAINTY 19 (1994). It bears noting that even supporters of health-health analysis acknowledge that the identified statistical correlation between income and health is far from a demonstration of causation. In fact, the argument linking regulatory expenditures to reduced health requires rather heroic faith in a series of counterfactual assumptions about the opportunity costs of regulation. *See* REVESZ & LIVERMORE, *supra* note 1, at 67–76; Richard W. Parker, *Grading the Government,* 70 U. CHI. L. REV. 1345, 1403 n.220 (2003).

10. *See* Douglas A. Kysar, *Some Realism about Environmental Skepticism: The Implications of Bjørn Lomborg's* The Skeptical Environmentalist *for Environmental Law and Policy,* 30 ECOLOGY L.Q. 223, 249–52 (2003).

11. *See* Daniel Bodansky, *The Precautionary Principle in U.S. Environmental Law,* in INTERPRETING THE PRECAUTIONARY PRINCIPLE 203 (Timothy O'Riordan & James Cameron eds., 1994).

12. "Wingspread Statement on the Precautionary Principle," *available at* http:// www.gdrc.org/u-gov/precaution-3.html.

13. Commission of the European Communities, COMMUNICATION FROM THE COMMISSION ON THE PRECAUTIONARY PRINCIPLE, Feb. 2, 2000, *available at* http://ec.europa.eu/dgs/health_consumer/library/pub/pub07_en.pdf. *See also* "Treaty Establishing the European Community," art. 130r.2, para. 1 ("[EU] policy on the environment shall . . . be based on the precautionary principle and on the principles that preventative action should be taken, that environmental damage should, as a priority, be rectified at source and that the polluter should pay.").

14. *Ethyl Corp. v. EPA,* 541 F.2d 1, 28 (D.C. Cir. 1976).

15. CASS R. SUNSTEIN, RISK & REASON: SAFETY, LAW, AND THE ENVIRONMENT 5–6 (2002). *See also* Robert W. Hahn & Cass R. Sunstein, *The Precautionary Principle as a Basis for Decision Making,* 2 ECONOMIST'S VOICE, no. 2, art. 8 (2005), at 6 ("We do not believe there is any principled way of making policy decisions without making the best possible effort to balance all the relevant costs of a policy against the benefits.").

16. DRAFT 2007 REPORT TO CONGRESS ON THE COSTS AND BENEFITS OF FEDERAL REGULATIONS 32, *available at* http://www.whitehouse.gov/omb/inforeg/2007_ cb/2007_draft_cb_report.pdf.

17. As one leading commentator observes, the wisdom of the precautionary principle is not necessarily to be found in its ability to provide fine-grained policy advice for particular risk dilemmas, but rather in its broader role as "part of a system of rules designed to guide human behavior towards the ideal of an environmentally sustainable economy." James Cameron, *The Precautionary Principle in International Law,* in REINTERPRETING THE PRECAUTIONARY PRINCIPLE 113, 113 (Tim O'Riordan, James Cameron & Andrew Jordan eds., 2001).

18. Cass R. Sunstein & Adrian Vermeule, *Is Capital Punishment Morally Required? Acts, Omissions, and Life-Life Tradeoffs*, 58 Stan. L. Rev. 703, 735 (2005).

19. For a formal demonstration of this point, *see* Robert C. Hockett, *The Impossibility of a Prescriptive Paretian*, Cornell Legal Studies Research Paper no. 06-027 (Sept. 2006), *available at* http://papers.ssrn.com/sol3/papers.cfm?abstract_id=930460.

20. Cass R. Sunstein, The Cost-Benefit State ix (2002) ("Gradually, and in fits and starts, American government is becoming a cost-benefit state.").

21. This phrase is an homage to Thomas Nagel's masterful work. *See* Thomas Nagel, The View from Nowhere (1986).

22. *Id.* at 26 (arguing that "any objective conception of reality must include an acknowledgment of its own incompleteness").

23. *Cf.* Clive L. Spash, Greenhouse Economics: Value and Ethics 244 (2002) (observing that "the Kyoto negotiations which are immersed in ethical considerations about resource distribution, rights and compensation have been treated as legal and economic discussions over technicalities").

24. Cass R. Sunstein, *Beyond the Precautionary Principle*, 151 U. Pa. L. Rev. 1003, 1008 (2003).

25. The term *ecological rationality* is associated with the work of psychologist Gerd Gigerenzer and refers to the notion that simple cognitive heuristics may represent "useful, even indispensable cognitive processes for solving problems that cannot be handled by logic and probability theory." Gerd Gigerenzer & Peter Todd, Simple Heuristics That Make Us Smart 25–26 (1999). For an overview, *see* Douglas A. Kysar et al., *Group Report: Are Heuristics a Problem or a Solution?* in Heuristics and the Law 103, 112 (Gerd Gigerenzer & Christoph Engel eds., 2006).

26. *See* Bruce Ackerman, *The Living Constitution*, 120 Harv. L. Rev. 1737 (2007). *See also* Bruce Ackerman, We the People: Foundations (1991); Bruce Ackerman, We the People: Transformations (1998).

27. These critiques are drawn, respectively, from Cass R. Sunstein, Worst-Case Scenarios 131 (2007); Todd J. Zywicki, *Baptists?: The Political Economy of Environmental Interest Groups*, 53 Case W. Res. L. Rev. 315, 333 (2002); Christopher D. Stone, *Is There a Precautionary Principle?* 31 Envtl. L. Rep. 10790, 10799 (2001); Sunstein, *supra* note 24, at 1004 (2003); Cass R. Sunstein, *Your Money or Your Life*, New Republic, Mar. 11, 2004, at 27; Sunstein, *Beyond the Precautionary Principle, supra* note 24, at 1008.

CHAPTER 1. AGENCY AND OPTIMALITY

1. *See* Judith Jarvis Thomson, *Killing, Letting Die, and the Trolley Problem*, 59 The Monist 204 (1976); Judith Jarvis Thomson, *The Trolley Problem*, 94 Yale L.J. 1395 (1985). Thompson builds on earlier work by Philippa Foot. *See* Philippa Foot, *The Problem of Abortion and the Doctrine of Double Effect*, in Moral

PROBLEMS: A COLLECTION OF PHILOSOPHICAL ESSAYS 59, 66 (James Rachels ed., 1975).

2. John Mikhail, *Universal Moral Grammar: Theory, Evidence, and the Future*, 11 TRENDS IN COGNITIVE SCI. 143 (2007).

3. *See* Foot, *supra* note 1, at 66 ("There is worked into our moral system a distinction between what we owe people in the form of aid and what we owe them in the way of non-interference.").

4. The Supreme Court of New Hampshire described this tradition through a particularly vivid example in *Buch v. Amory Mfg. Co.*, 44 A. 809, 811 (N.H. 1897): "I see my neighbor's two-year-old babe in dangerous proximity to the machinery of his windmill in his yard, and easily might, but do not, rescue him. I am not liable in damages to the child for his injuries . . . because the child and I are strangers, and am under no legal duty to protect him."

5. *See* Ilana Ritov & Jonathan Baron, *Reluctance to Vaccinate: Omission Bias and Ambiguity*, in BEHAVIORAL LAW AND ECONOMICS 168, 184 (Cass R. Sunstein ed., 2002) (reporting that "[s]ubjects are reluctant to vaccinate when the vaccine can cause bad outcomes, even if the outcomes of not vaccinating are worse").

6. *See* Robert A. Prentice & Jonathan J. Koehler, *A Normality Bias in Legal Decision Making*, 88 CORNELL L. REV. 583, 587 (2003) (describing the "omission bias" as "the tendency of people to find more blameworthy bad results that stem from actions than bad results that stem from otherwise equivalent omissions"); Laura Y. Niedermayer & Gretchen B. Chapman, *Action, Inaction, and Factors Influencing Perceived Decision Making*, 14 J. BEHAV. DECISION MAKING 295, 296 (2001) (defining the bias as "the tendency to judge actions as worse than omissions when they both have the same bad consequences"); Jonathan Baron, *Nonconsequentialist Decisions*, 17 BEHAV. & BRAIN SCI. 1, 3 (1994) (observing that "[p]eople continue to distinguish acts and omissions . . . even when the feature that typically makes them different is absent").

7. *See* JONATHAN BARON, JUDGMENT MISGUIDED: INTUITION AND ERROR IN PUBLIC DECISION MAKING 196–99 (1998) (calling for education beginning in elementary school regarding utilitarian thinking); Jonathan Baron, *Heuristics and Biases in Equity Judgments*, in PSYCHOLOGICAL PERSPECTIVES ON JUSTICE 109, 135 (Barbara A. Mellers & Jonathan Baron eds., 1993) (noting that "people should be taught to understand the utilitarian approach").

8. WERNER HEISENBERG, PHYSICS AND PHILOSOPHY 96 (1958).

9. *Cf.* THOMAS W. POGGE, WORLD POVERTY AND HUMAN RIGHTS: COSMOPOLITAN RESPONSIBILITIES AND REFORMS (2002) (emphasizing the affirmative role of developed nations in supporting a global order that contributes to and maintains poverty).

10. *See, e.g.*, RICHARD B. BRANDT, A THEORY OF THE GOOD AND THE RIGHT (1979).

11. *See* Cass R. Sunstein, *Moral Heuristics and Moral Framing*, 88 MINN. L. REV. 1556 (2004).

12. Bernard Williams, *A Critique of Utilitarianism*, in ETHICS: HISTORY, THEORY, AND CONTEMPORARY ISSUES 567, 582 (Steven M. Cahn & Peter Markie eds., 1998).

13. Bernard Williams, *A Critique of Utilitarianism*, in J. J. C. SMART & BERNARD WILLIAMS, UTILITARIANISM: FOR AND AGAINST 116–17 (1973).

14. *See* SAMUEL SCHEFFLER, BOUNDARIES AND ALLEGIANCES: PROBLEMS OF JUSTICE AND RESPONSIBILITY IN LIBERAL THOUGHT 121 (2001) (arguing that "interpersonal relationships cannot play the fundamental role that they do in human life unless people treat their own relationships as independent sources of reasons for action").

15. *See* THOMAS NAGEL, THE VIEW FROM NOWHERE (1986).

16. *See* John Greenleaf Whittier, *Maud Muller*, in ENGLISH POETRY III: FROM TENNYSON TO WHITMAN (Charles W. Gliot ed., 1909) ("For of all sad words of tongue or pen, / The saddest are these: 'It might have been!' ").

17. *See* Williams, *supra* note 13, at 94 (noting that "for consequentialism, all causal connexions are on the same level, and it makes no difference, so far as that goes, whether the causation of a given state of affairs lies through another agent, or not").

18. Joshua D. Greene, *The Secret Joke of Kant's Soul*, in 3 MORAL PSYCHOLOGY: THE NEUROSCIENCE OF MORALITY: EMOTION, DISEASE, AND DEVELOPMENT 35, 36 (Walter Sinnott-Armstrong ed., 2008). *See also* Jonathan Haidt, *The Emotional Dog and Its Rational Tail: A Social Intuitionist Approach to Moral Judgment*, 108 PSYCHOL. REV. 814 (2001) (proposing a "social intuitionist" model of moral decision making, in which judgments of right and wrong are seen to flow from a process of intuitive recognition of moral emotions that have evolved over time to serve certain valuable interpersonal functions).

19. Greene, *supra* note 18, at 37–38.

20. *Id.* at 39.

21. *See* Michael Koenigs et al., *Damage to the Prefrontal Cortex Increases Utilitarian Moral Judgements*, 446 NATURE 908 (2007).

22. *Id.* at 909–10.

23. Greene, *supra* note 18, at 46.

24. *Id.* at 60.

25. Sunstein, *supra* note 11, at 1558.

26. *Id.* at 1580.

27. *See* Jedediah Purdy, *The Promise (and Limits) of Neuroeconomics*, 58 ALA. L. REV. I, 21–40 (2006).

28. *See* JOHN D. BARROW, IMPOSSIBILITY: THE LIMITS OF SCIENCE AND THE SCIENCE OF LIMITS 91 (1998) ("Ironically, the most complicated thing we have encountered in the entire panorama of Nature, from the inner space of the elementary particles of matter to the outer space of distant galaxies, is what lies inside our heads").

29. Benedict Carey, *Study Finds Brain Injury Changes Moral Judgment*, N.Y. TIMES, Mar. 21, 2007, *available at* http://www.nytimes.com/2007/03/21/health/ 21cnd-brain.html.

30. For an overview of a highly fruitful field of psychological research exploring cultural dimensions of cognition, *see* Dan M. Kahan & Donald Braman, *Cultural Cognition and Public Policy*, 24 YALE L. & POL'Y REV. 149 (2006).

31. *Cf.* THOMAS NAGEL, MORTAL QUESTIONS 210 (1979) (referring to the problems of free will, personal identity, and the mind-body as examples of what Nietzsche called "[t]he indigestible lump" that gives pause to the totalizing ambitions of objective views).

32. Samual Scheffler, *Doing and Allowing*, 114 ETHICS 215, 232 (2004).

33. *Id.*

34. Daniel Markovits, *Legal Ethics and the Lawyer's Point of View*, 15 YALE J. L. & HUMAN. 209, 249 (2003). *See also* JED RUBENFELD, FREEDOM AND TIME: A THEORY OF CONSTITUTIONAL SELF-GOVERNMENT 97 (2001) ("The committed self wants to be governed not by his present will or voice, but, in important part, by texts of his own authorship, whether or not these texts are literally written down").

35. NAGEL, *supra* note 15, at 152–53.

36. SAMUEL SCHEFFLER, THE REJECTION OF CONSEQUENTIALISM 94 (1995).

CHAPTER 2. PRESCRIPTION AND PRECAUTION

1. *See* CLIVE L. SPASH, GREENHOUSE ECONOMICS: VALUE AND ETHICS 132 (2002).

2. Admittedly, this is an unduly simplified account. For a more complete rendering, including discussion of the significant role of U.S. economic interests in a global ozone regime, *see* RICHARD ELLIOTT BENEDICK, OZONE DIPLOMACY: NEW DIRECTIONS IN SAFEGUARDING THE PLANET (1991).

3. *See* John D. Graham, *The Role of Precaution in Risk Assessment and Management: An American's View* (2002), http://www.whitehouse.gov/omb/inforeg/ eu_speech.html.

4. Cass R. Sunstein, *Beyond the Precautionary Principle*, 151 U. PA. L. REV. 1003, 1008 (2003).

5. For important contributions and discussions, *see* KENNETH J. ARROW, SOCIAL CHOICE AND INDIVIDUAL VALUES (2d ed. 1963); AMARTYA K. SEN, COLLECTIVE CHOICE AND SOCIAL WELFARE (1970); Kenneth J. Arrow, *A Difficulty in the Concept of Social Welfare*, 58 J. POL. ECON. 328 (1950); Kenneth J. Arrow, *Some Ordinalist-Utilitarian Notes on Rawls's Sense of Justice*, 70 J. PHIL. 245 (1973); Bruce Chapman, *More Easily Done Than Said: Rules, Reasons and Rational Social Choice*, 18 OXFORD J. LEGAL STUD. 293 (1998); Lewis A. Kornhauser & Lawrence G. Sager, *The Many as One: Integrity and Group Choice in Paradoxical Cases*, 32 PHIL. & PUB. AFF. 249 (2004); Charles R. Plott, *Axiomatic Social Choice Theory: An Overview and Interpretation*, 20 AM. J. POL. SCI. 511 (1976).

6. The reductionist view of the state within international relations theory—in which the state is merely understood as an assemblage of its individual members—shares both the exclusively individualistic metaphysics of social choice theory and the underlying fear of nonliberal politics. *See* Alexander Wendt, *The State as a Person in International Theory*, 30 REV. INT'L STUD. 289, 315 (2004).

7. *See* JOHN RAWLS, A THEORY OF JUSTICE (1971); ROBERT NOZICK, ANARCHY, STATE, AND UTOPIA (1974); BRUCE A. ACKERMAN, SOCIAL JUSTICE IN THE LIBERAL STATE (1980).

8. *See* ROBERT NOZICK, ANARCHY, STATE, AND UTOPIA (1974).

9. Indeed, some contend that libertarianism so dramatically emphasizes private ordering at the cost of a public sphere as to fail to count as a liberal theory. *See* Samuel Freeman, *Illiberal Libertarians: Why Libertarianism Is Not a Liberal View*, 30 PHIL. & PUB. AFF. 105 (2001).

10. *Cf.* Richard Craswell, *Incommensurability, Welfare Economics, and the Law*, 146 U. PA. L. REV. 1419, 1461 (1998) ("[B]y most liberal accounts the government is an inanimate institution, which is justified (if at all) by what it contributes to its individual citizens.").

11. *See, e.g.,* MICHAEL WALZER, SPHERES OF JUSTICE: A DEFENSE OF PLURALISM AND EQUALITY (1983); MICHAEL SANDEL, LIBERALISM AND THE LIMITS OF JUSTICE (1982); ALASDAIR MACINTYRE, AFTER VIRTUE: A STUDY IN MORAL THEORY (1981).

12. The debate is ably synthesized in RAINER FORST, CONTEXTS OF JUSTICE: POLITICAL PHILOSOPHY BEYOND LIBERALISM AND COMMUNITARIANISM (2002).

13. *See* John Rawls, *Justice as Fairness: Political Not Metaphysical*, 14 PHIL. & PUB. AFFAIRS 223 (1985).

14. Jürgen Habermas, *Remarks on Legitimation Through Human Rights*, 24 PHIL. & SOC. CRITICISM 157, 167 (1998).

15. This identity or agency can often be detected within communitarian descriptions of the significance of the state to individual identity. *See, e.g.,* WILL KYMLICKA, CONTEMPORARY POLITICAL PHILOSOPHY: AN INTRODUCTION 264 (2d ed. 2002) ("The basis of a common national identity need not be a shared conception of the good, but rather a thinner and more diffuse sense of belonging to an intergenerational society, sharing a common territory, having a common past and sharing a common future."); Alasdair MacIntyre, *Is Patriotism a Virtue?* in THEORISING CITIZENSHIP 209, 224 (Ronald Beiner ed., 1995) ("A central contention of the morality of patriotism is that I will obliterate and lose a central dimension of the moral life if I do not understand the enacted narrative of my own individual life as embedded in the history of my country.").

16. *See* ROBERT E. GOODIN, UTILITARIANISM AS A PUBLIC PHILOSOPHY 27 (1995) (concluding that "[t]he same thing that makes [moral] excuses valid at the individual level—the same thing that relieves individuals of responsibility [from a duty of causal optimality]—makes it morally incumbent upon individuals to

organize themselves into collective units that are capable of acting where they as isolated individuals are not"); Thomas Nagel, *Ruthlessness in Public Life,* in PUBLIC and PRIVATE MORALITY 75, 84 (Stuart Hampshire ed., 1978) ("Within the appropriate limits, public decisions will be justifiably more consequentialist than private ones.").

17. *See generally* BRUCE ACKERMAN, WE THE PEOPLE: FOUNDATIONS (1991).

18. Eric A. Posner & Cass R. Sunstein, *Climate Change Justice,* 96 GEO. L.J. 1565, 1572 (2008).

19. *See* Patrick Thaddeus Jackson, *Is the State a Person? Why Should We Care?* 30 REV. INT'L STUD. 255 (2004) (summarizing recent literature).

20. Hegel believed that the state "must be apprehended as an organism," but one that exists between atomistic individualism and totalitarianism. *See* ROBERT R. WILLIAMS, HEGEL'S ETHICS OF RECOGNITION 294–95 (1997).

21. *See* JED RUBENFELD, FREEDOM AND TIME: A THEORY OF CONSTITUTIONAL SELF-GOVERNMENT (2001).

22. ALEXANDER WENDT, SOCIAL THEORY OF INTERNATIONAL POLITICS 194 (1999).

23. Patrick Thaddeus Jackson, *Hegel's House, or 'People Are States Too,'* 30 REV. INT'L STUD. 281, 281 (2004).

24. As Thomas Nagel has written, "Even though the morality of politics is rightly more impersonal than the morality of private life, the acknowledgment of personal values and autonomy is essential even at the level that requires the greatest impersonality." THOMAS NAGEL, THE VIEW FROM NOWHERE 188 (1986).

25. Cass R. Sunstein & Adrian Vermeule, *Is Capital Punishment Morally Required? Acts, Omissions, and Life-Life Tradeoffs,* 58 STAN. L. REV. 703, 707 (2005).

26. *See* John J. Donohue & Justin Wolfers, *Uses and Abuses of Empirical Evidence in the Death Penalty Debate,* 58 STAN. L. REV. 791 (2005).

27. Sunstein & Vermeule, *supra* note 25, at 707.

28. *Id.* at 708 (emphasis added).

29. *Id.* at 706.

30. *Id.* at 708.

31. *See id.* at 709 ("If capital punishment has significant deterrent effects, we suggest that for government to omit to impose it is morally blameworthy, even on a deontological account of morality. Deontological accounts typically recognize a consequentialist override to baseline prohibitions. If each execution saves an average of eighteen lives, then it is plausible to think that the override is triggered, in turn triggering an obligation to adopt capital punishment.").

32. *Id.* at 708.

33. *Id.* at 710, 734, 735.

34. *Id.* at 728.

35. *See* Mark Kelman, *Taking Takings Seriously: An Essay for Centrists,* 74 CAL. L. REV. 1829, 1849–51 (1986) (reviewing RICHARD A. EPSTEIN, TAKINGS:

PRIVATE PROPERTY AND THE POWER OF EMINENT DOMAIN (1985) and arguing that to translate individual rights and responsibilities uncritically to the collective sphere is to "reify like mad"); Foreword to PUBLIC and PRIVATE MORALITY, *supra* note 16 (collecting papers by Stuart Hampshire, T. M. Scanlon, Bernard Williams, Thomas Nagel, and Ronald Dworkin regarding "the dividing line between private life and public responsibilities").

36. ROBERT E. GOODIN, UTILITARIANISM AS A PUBLIC PHILOSOPHY 9 (1995) ("It is the essence of public service as such that public servants should serve the public at large."); Nagel, *supra* note 16, at 83 ("Public institutions are designed to serve purposes larger than those of particular individuals or families.").

37. *See* SAMUEL SCHEFFLER, BOUNDARIES AND ALLEGIANCES: PROBLEMS OF JUSTICE AND RESPONSIBILITY IN LIBERAL THOUGHT 43 (2001) (noting that "the individual agent *qua* individual agent typically will have only the most limited opportunities to influence . . . global dynamics, and, indeed, cannot in general be assumed to have any but the sketchiest and most speculative notions about the specific global implications of his or her personal behavior"); Richard Trammell, *Saving Life and Taking Life,* 72 J. PHIL. 131, 133 (1975) (using a notion of "dischargeability," which asks whether a proposed normative standard could ever be satisfied, to argue against a utilitarian duty to rescue).

38. *See* Kelman, *supra* note 35, at 1851 (noting that states, as opposed to individuals, can more readily formulate "the scope of [positive] duties in administrable rule-like form" and utilize "a general system of taxation" to distribute the costs of fulfilling such duties in an equitable manner).

39. Michael J. Green, *Institutional Responsibility for Global Problems,* 30 PHIL. TOPICS 79, 86 (2002).

40. *See* GOODIN, *supra* note 36, at 129 (arguing that "public officials take a longer time horizon than do individuals planning their own private lives"); SCHEFFLER, *supra* note 37, at 39 (noting that individuals "tend to experience . . . causal influence as inversely related to spatial and temporal distance").

41. Green, *supra* note 39, at 86.

42. Sunstein & Vermeule, *supra* note 25, at 706.

43. *Cf.* Bernard Williams, *A Critique of Utilitarianism,* in J. J. C. SMART & BERNARD WILLIAMS, UTILITARIANISM FOR AND AGAINST 112 (1973) ("Utilitarianism has a tendency to slide in this direction, and to leave a vast hole in the range of human desires, between egoistic inclinations and necessities at one end, and impersonally benevolent happiness-management at the other.").

44. Sunstein & Vermeule, *supra* note 25, at 709.

45. *Id.* at 720.

46. Scott Shane & Eric Lipton, *Government Saw Flood Risk but Not Levee Failure,* N.Y. TIMES, Sept. 2, 2005, at A1 (quoting a White House official "who asked not to be named because he did not want to be seen as talking about the crisis in political terms").

47. George W. Bush, "Address to Volunteers at Evacuee Shelter in Baton Rouge, Louisiana" Sept. 5, 2005), transcript *available at* http://www.whitehouse.gov/news/releases/2005/09/20050905-9.html.

48. Sunstein & Vermuele, *supra* note 25, at 721.

49. *Id.* at 714 n.110.

50. *Id.* at 750.

51. Cass R. Sunstein & Adrian Vermeule, *Deterring Murder: A Reply,* 58 STAN. L. REV. 847, 851 (2005).

52. Sunstein & Vermuele, *supra* note 25, at 725 ("Government officials cannot plausibly claim that their liberty is abridged when citizens ask them to take steps against, say, domestic violence, occupational deaths, or rape."). *See also* J. A. Mirrlees, *The Economic Uses of Utilitarianism,* in UTILITARIANISM AND BEYOND 63, 71 (Amartya Sen & Bernard Williams eds., 1982) (arguing that "[i]n the evaluation of the outcomes of public policy, loyalty and other kinds of partiality should be excluded," but construing partiality only to refer to individual government agents, rather than to the political entity itself).

53. S. Boehmer-Christiansen, *The Precautionary Principle in Germany—Enabling Government,* in INTERPRETING THE PRECAUTIONARY PRINCIPLE 34, 38 (Timothy O'Riordan & James Cameron eds., 1994) (emphasis added).

54. Critics often believe that adherence to the precautionary principle suggests that individuals naïvely regard the status quo, the nonhuman, or the "normal" causal order as benign. *See* Sunstein, *supra* note 4, at 1009 (arguing that the "mistaken belief that nature is essentially benign . . . often informs the precautionary principle"). Although most precautionary principle proponents harbor no such illusion, the more fundamental point is that *some* counterfactual baseline (such as the "normal" causal order, absent one's actions) is necessary in order for any form of consequentialist moral reasoning about human behavior to coherently proceed. After all, as chapter 4 argues, cost-benefit analysis has a baseline of its own, premised on a market-liberal conception in which existing economic arrangements and preferences are given privileged status.

55. Michael Moore's contrary argument—that human omissions are "literally nothing at all"—seems just as difficult to sustain as the utilitarian notion that human actions are indistinguishable from other causal forces. MICHAEL MOORE, ACT AND CRIME: THE PHILOSOPHY OF ACTION AND ITS IMPLICATIONS FOR CRIMINAL LAW 28 (1993). In both cases, the critical missing element is an appreciation for the role played by human choice and agency. *See* George P. Fletcher, *On the Moral Irrelevance of Bodily Movements,* 142 U. PA. L. REV. 1443, 1444 (1994) (arguing that "the only kind of omitting that is interesting is the kind in which human agency is expressed").

56. *See* A. Dan Tarlock, *The Nonequilibrium Paradigm in Ecology and the Partial Unraveling of Environmental Law,* 27 LOY. L.A. L. REV. 1121 (1994).

57. NAGEL, *supra* note 24, at 209.

58. *See* Amartya Sen & Bernard Williams, *Introduction: Utilitarianism and Beyond*, in UTILITARIANISM AND BEYOND, *supra* note 52, at 16 (stating that "Government House utilitarianism," such as the approach embodied in cost-benefit-analysis risks becoming "an outlook favouring social arrangements under which a utilitarian elite controls a society in which the majority may not itself share those beliefs"). Sen and Williams refer particularly to the partial-disclosure conception of utilitarianism, in which elites apply the utilitarian calculus to social decision making without disclosing their method of analysis, given the fear that its cold calculation might undermine the basis of social cohesion among citizens less capable of such supposedly enlightened reasoning.

CHAPTER 3. COMPLEXITY AND CATASTROPHE

1. Cass R. Sunstein & Adrian Vermeule, *Is Capital Punishment Morally Required? Acts, Omissions, and Life-Life Tradeoffs*, 58 STAN. L. REV. 703, 723 (2005).

2. J. B. Ruhl, *Sustainable Development: A Five-Dimensional Algorithm for Environmental Law*, 18 STAN. ENVTL. L.J. 31, 46 (1999).

3. Adapted from Andy Stirling, *The Precautionary Principle in Science and Technology*, in REINTERPRETING THE PRECAUTIONARY PRINCIPLE 61, 79 (Tim O'Riordan, James Cameron & Andrew Jordan eds., 2001); and Andy Stirling, Ortwin Renn & Patrick van Zwanenberg, *A Framework for the Precautionary Governance of Food Safety: Integrating Science and Participation in the Social Appraisal of Risk*, in IMPLEMENTING THE PRECAUTIONARY PRINCIPLE: PERSPECTIVES AND PROSPECTS 284, 288 (Elizabeth Fisher, Judith Jones & René von Schomberg eds., 2006).

4. *See* Daniel A. Farber, *Probabilities Behaving Badly: Complexity Theory and Environmental Uncertainty*, 37 U.C. DAVIS L. REV. 145, 152–55 (2003).

5. J. B. Ruhl, *Complexity Theory as a Paradigm for the Dynamical Law-and-Society System: A Wake-Up Call for Legal Reductionism and the Modern Administrative State*, 45 DUKE L.J. 849, 937 (1996).

6. *See* JAMES GLEICK, CHAOS: MAKING A NEW SCIENCE 22 (1987).

7. This section draws heavily on Douglas A. Kysar & Thomas O. McGarity, *Did NEPA Drown New Orleans? The Levees, the Blame Game, and the Hazards of Hindsight*, 56 DUKE L.J. 179 (2006).

8. HOWARD E. GRAHAM & DWIGHT E. NUNN, DEP'T OF COMMERCE, NAT'L HURRICANE RESEARCH PROJECT REPORT NO. 33, METEOROLOGICAL CONSIDERATIONS PERTINENT TO STANDARD PROJECT HURRICANE, ATLANTIC AND GULF COASTS OF THE UNITED STATES 1 (1959). For a nontechnical overview of the standard project hurricane and related engineering issues, *see* J. J. Westerink & R. A. Leuttich, *The Creeping Storm*, CIV. ENGINEERING MAG., June 2003, at 46, 48–52.

9. *Flood Control Act of 1936*, ch. 688, § 1, 49 Stat. 1570, 1570 (codified as amended at 33 U.S.C. § 701(a) (2000)).

10. *See* GRAHAM & NUNN, *supra* note 8, at 1. ("The standard project hurricane wind field and parameters represent a 'standard' against which the degree of protection finally selected for a hurricane protection project may be judged and compared with protection provided at projects in other localities.")

11. These quotes are drawn, respectively, from HYDROMETEOROLOGICAL BRANCH, NAT'L OCEANIC & ATMOSPHERIC ADMIN., MEMORANDUM HUR 7-120, PRELIMINARY REVISED STANDARD PROJECT HURRICANE CRITERIA FOR THE ATLANTIC AND GULF COASTS OF THE UNITED STATES 3 (1972); U.S. ARMY ENG'R DIST., LAKE PONTCHARTRAIN, LOUISIANA, AND VICINITY HURRICANE PROTECTION PLAN I-2 (1974); *id.* at VIII-5; *id.* at VIII-11; and John McQuaid & Bill Walsh, *Warnings to Beef up New Orleans' '60s-Era Levees Unheeded,* NEWHOUSE NEWS SERVICE, Sept. 2, 2005, *available at* www.katrina.eng.lsu.edu/pdf_articles/suhayda_newhousenews.pdf (quoting Gen. Carl Strock, Army Corps Chief of Engineers).

12. Daniel A. Farber, *Probabilities Behaving Badly: Complexity Theory and Environmental Uncertainty,* 37 U.C. DAVIS L. REV. 145 (2003).

13. Kysar & McGarity, *supra* note 7, at 221.

14. *See* U.S. ARMY CORPS OF ENGINEERS, PERFORMANCE EVALUATION OF THE NEW ORLEANS AND SOUTHEAST LOUISIANA HURRICANE PROTECTION SYSTEM, DRAFT FINAL REPORT OF THE INTERAGENCY PERFORMANCE EVALUATION TASK FORCE, VOLUME VIII: ENGINEERING AND OPERATIONAL RISK AND RELIABILITY ANALYSIS, June 1, 2006, at VIII-37 (2006) ("[H]urricane rates are uncertain due to the historical sample size, possible errors in the assumed form of marginal and conditional distributions (especially in the tail regions), and the uncertain near-future hurricane activity due to fluctuations and trends associated with climate changes and multi-decadal cycles.").

15. GRAHAM & NUNN, *supra* note 8, at 12.

16. *See* U.S. ARMY CORPS OF ENGINEERS, PERFORMANCE EVALUATION OF THE NEW ORLEANS AND SOUTHEAST LOUISIANA HURRICANE PROTECTION SYSTEM, DRAFT FINAL REPORT OF THE INTERAGENCY PERFORMANCE EVALUATION TASK FORCE, VOLUME I: EXECUTIVE SUMMARY AND OVERVIEW, at I-26 (2007) (stating that SPH was meant to capture a "fast-moving Category 3 hurricane" and that for New Orleans this translated to a "200- to 300-year recurrence event"); U.S. GOV'T ACCOUNTABILITY OFFICE, HURRICANE PROTECTION: STATUTORY AND REGULATORY FRAMEWORK FOR LEVEE MAINTENANCE AND EMERGENCY RESPONSE FOR THE LAKE PONTCHARTRAIN PROJECT 4 (2005) ("[The SPH] was expected to have a frequency of occurrence of once in about 200 years, and represented the most severe combination of meteorological conditions considered reasonably characteristic for the region.").

17. HYDROMETEOROLOGICAL BRANCH, *supra* note 11, at 3.

18. *See, e.g.,* U.S. GOV'T ACCOUNTABILITY OFFICE, *supra* note 16, at 4.

19. HYDROMETEOROLOGICAL BRANCH, *supra* note 11, at 3.

20. Harry S. Perdikis, *Hurricane Flood Protection in the United States,* 93 J. WATERWAYS & HARBORS DIVISION, Feb. 1967, at 9. Even on narrow economic

grounds, though, the choice of a one-hundred-year return period for natural-disaster planning might be questioned: Studies suggest, for instance, that a large majority (66–83 percent) of losses from floods and hurricane winds come from events with recurrence intervals less frequent than the one-hundred-year flood. *See* Raymond J. Burby, *Hurricane Katrina and the Paradoxes of Government Disaster Policy: Bringing about Wise Governmental Decisions for Hazardous Areas,* ANNALS AM. ACAD. POL. & SOC. SCI., Mar. 2006, at 171, 177.

21. *See Hurricane Protection Plan for Lake Pontchartrain and Vicinity: Hearing Before the House Subcommittee on Water Resources,* 95th Cong. 16 (1978) (statement of Col. Early J. Rush III, District Engineer, U.S. Army Engineer District, New Orleans: "Even though economists may, and in this case, did, favor protection to a lower scale to produce a higher ratio of project benefits to project costs, the threat of loss of human life mandated using the standard project hurricane."); SELECT BIPARTISAN COMM. TO INVESTIGATE THE PREPARATION for and RESPONSE TO HURRICANE KATRINA, A FAILURE OF INITIATIVE 89–90 (2006) (quoting testimony).

22. RICHARD W. SCHWERDT ET AL., NAT'L OCEANIC & ATMOSPHERIC ADMIN., NOAA TECHNICAL REPORT NWS 23, METEOROLOGICAL CRITERIA FOR STANDARD PROJECT HURRICANE AND PROBABLE MAXIMUM HURRICANE WIND FIELDS, GULF AND EAST COASTS OF THE UNITED STATES 143 (1979). *See also id.* ("These two hurricanes are much more severe than any other in the gulf and are therefore not 'reasonably characteristic.'"); *id.* at 2. ("By reasonably characteristic is meant that only a few hurricanes of record over a large region have had more extreme values of the meteorological parameters.")

23. *Id.* at 5.

24. Perdikis, *supra* note 20, at 9.

25. U.S. GOV'T ACCOUNTABILITY OFFICE, CORPS OF ENGINEERS: IMPROVED ANALYSIS OF COSTS AND BENEFITS NEEDED FOR SACRAMENTO FLOOD PROTECTION PROJECT 20 n.13 (2003).

26. *See* U.S. ARMY ENG'R DIST., *supra* note 11, at VIII-12 (1974).

27. *See id.* at title page (reporting only property damage prevention, land intensification, and redevelopment as itemized annual benefits); *id.* at VIII-21. ("Environmental losses were not evaluated in dollar terms.")

28. *See id.* at ii ("Indirectly, the plan will hasten urbanization and industrialization of valuable marshes and swamps by providing for further flood protection and land reclamation."); *see also id.* at VIII-27 ("Several areas would be rendered more suitable for urban use as a result of the project works. This effect will be reflected in increases in value of these lands, which increases are called 'enhancement benefits,' since they do represent additions to the Gross National Product.").

29. Burby, *supra* note 20, at 174 (citing U.S. GEN. ACCOUNTING OFFICE, COST, SCHEDULE, AND PERFORMANCE PROBLEMS OF THE LAKE PONTCHARTRAIN AND VICINITY, LOUISIANA, HURRICANE PROTECTION PROJECT 3 (1976)). This con-

flation of protection and promotion purposes appears to be common within flood-control and hurricane-protection planning. *See* Raymond J. Burby & Steven P. French, Flood Plain Land Use Management: A National Assessment 146–47 (1985) (finding a positive correlation between community flood-control works and the amount of new development taking place in flood-hazard areas after flood-control works are completed).

30. *See* Kysar & McGarity, *supra* note 7, at 191–98.

31. U.S. Army Corps of Engineers, *supra* note 16, at I-64.

32. *Id.* at I-4.

33. *Id.* at I-72.

34. As cognitive psychologist Gerd Gigerenzer has shown, individuals appear to process risk information much more reliably when it is presented in frequency rather than probability terms. *See, e.g.,* Gerd Gigerenzer, *The Bounded Rationality of Probabilistic Mental Models,* in Rationality: Psychological and Philosophical Perspectives 284 (K. I. Manktelow & D. E. Over eds., 1993); Gerd Gigerenzer, *Ecological Intelligence: An Adaptation for Frequencies,* in The Evolution of Mind 9, 11–15 (Denise Dellarosa Cummins & Colin Allen eds., 1998); Gerd Gigerenzer & Ulrich Hoffrage, *How to Improve Bayesian Reasoning without Instruction: Frequency Formats,* 102 Psychol. Rev. 684, 697–98 (1995).

35. As the U.S. General Accounting Office reported in 1982, state and local sponsors in New Orleans repeatedly "recommended that the Corps *lower* its design standards to provide more realistic hurricane protection to withstand a hurricane whose intensity might occur once every 100 years rather than building a project to withstand a once in 200- to 300-year occurrence." U.S. Gen. Accounting Office, Improved Planning Needed by the Corps of Engineers to Resolve Environmental, Technical, and Financial Issues on the Lake Pontchartrain Hurricane Protection Project app. I at 9 (1982) (emphasis added). This hesitancy surely arose because state and local officials knew their agencies would have to share the costs of an expanded LPVHPP project with the federal government under the enabling statute for Army Corps flood- and hurricane-protection projects.

36. John McQuaid & Mark Schleifstein, Path of Destruction: The Devastation of New Orleans and the Coming Age of Superstorms 361 (2006).

37. *See* Douglas A. Kysar, *Preferences for Processes: The Process/Product Distinction and the Regulation of Consumer Choice,* 118 Harv. L. Rev. 525 (2004).

38. Wendy E. Wagner, *The Precautionary Principle and Chemical Regulation in the U.S.,* 6 Hum. & Ecological Risk Assessment 459, 466–68 (2000).

39. The assumed equivalency approach to assessing nanoscale substances has not been limited to the FDA. *See* J. Clarence Davies, Woodrow Wilson Int'l Ctr. for Scholars, Project on Emerging Nanotechnologies, EPA and Nanotechnology: Oversight for the 21st Century 31 (PEN 9, 2007), *available at* http://www.nanotechproject.org/process/assets/files/ 2698/197_nanoepa_pen9.pdf (observing that "[u]nder questioning from a

reporter" EPA revealed that only one of fifteen new nanoscale chemicals had been found to have "unique properties that would cause it to act differently than a larger form of the same chemical," but the EPA failed to reveal "how it defined 'unique properties' and did not indicate what evidence, if any, it used to reach its conclusion").

40. *See* Kenneth J. Arrow & Anthony C. Fisher, *Environmental Preservation, Uncertainty, and Irreversibility,* 88 Q. J. ECON. 312 (1974). For a leading efficiency model incorporating option values, *see* A. Myrick Freeman, *The Sign and Size of Option Value,* 60 LAND ECON. 1 (1984).

41. *See* Arrow & Fisher, *supra* note 40, at 317.

42. *Id.* at 319.

43. *See* CASS R. SUNSTEIN, WORST-CASE SCENARIOS 166 (2007).

44. *See* M. L. Weitzman, "Structural Uncertainty and the Value of Statistical Life in the Economics of Catastrophic Climate Change" (Nat'l Bureau of Econ. Research, Working Paper No. W13490, 2007), *available at* http://papers.ssrn.com/sol3/papers.cfm?abstract_id=1021968 (arguing that expected value calculation, even when performed with sophisticated Monte Carlo techniques, may radically distort decision making, since those techniques tend to assume a normal or lognormal distribution).

45. For discussion of Bayesian probability theory, *see* David E. Adelman, *Scientific Activism and Restraint: The Interplay of Statistics, Judgment, and Procedure in Environmental Law,* 79 NOTRE DAME L. REV. 497 (2004); Matthew D. Adler, *Against "Individual Risk": A Sympathetic Critique of Risk Regulation,* 153 U. PA. L. REV. 1121 (2005); Stephen Charest, *Bayesian Approaches to the Precautionary Principle,* 12 DUKE ENVTL. L. & POL'Y FORUM 265 (2002). As Charest points out, the Bayesian approach may be an improvement over risk-assessment techniques that incorporate equally subjective assumptions through less transparent means.

46. *See* SUNSTEIN, *supra* note 43, at 159 (suggesting that uncertainty does not exist as a distinct epistemic category since "personal probability" always is available).

47. In 2002, a specially appointed committee of the National Research Council concluded that "greenhouse warming and other human alterations of the earth system may increase the possibility of large, abrupt, and unwelcome regional or global climatic events." COMM. ON ABRUPT CLIMATE CHANGE OF THE NAT'L RESEARCH COUNCIL, ABRUPT CLIMATE CHANGE: INEVITABLE SURPRISES 1 (2002); *see also* R. B. Alley et al., *Abrupt Climate Change,* 299 SCI. 2005 (2003).

48. Martin L. Weitzman, *A Review of* The Stern Review on the Economics of Climate Change, 45 J. ECON. LITERATURE 703, 716 (2007).

49. *Id.*

50. *See* CONG. BUDGET OFFICE, UNCERTAINTY IN ANALYZING CLIMATE CHANGE: POLICY IMPLICATIONS 16 (2005) (observing that, with respect to the survey relied on by Nordhaus and Boyer, "[t]he experts' damage estimates seem to depend on their particular area of knowledge: natural scientists tended to

predict larger losses than social scientists did, with the bulk of the expected losses consisting of damages to non-marketed goods and services rather than the types of goods and services that are measured in standard national economic accounts").

51. WILLIAM NORDHAUS & JOSEPH BOYER, WARMING THE WORLD: ECONOMIC MODELING OF GLOBAL WARMING 87 (2000).

52. *Id.* at 88.

53. CLIVE L. SPASH, GREENHOUSE ECONOMICS: VALUE AND ETHICS 127 (2002).

54. Thomas O. McGarity, *A Cost-Benefit State*, 50 ADMIN. L. REV. 7 (1998).

55. Cass R. Sunstein, *Valuing Life: A Plea for Disaggregation*, 54 DUKE L.J. 385, 433 (2004).

56. *See* CASS R. SUNSTEIN, RISK AND REASON: SAFETY, LAW, AND THE ENVIRONMENT 108 (2002). Sunstein's analysis of the costs and benefits of arsenic regulation, for instance, noted that plausible assumptions could support benefit estimates from $10 million to $1.2 billion for a proposed standard. *See* Cass R. Sunstein, *The Arithmetic of Arsenic*, 90 GEO. L.J. 2255, 2258 (2002).

57. SPASH, *supra* note 53, at 115 ("In order to study the effects of changes in atmospheric chemistry upon biological systems, data analysis must be based upon fluxes, variances, extreme events and 'noise,' rather than concentrations, smoothed means and steady states.").

CHAPTER 4. INTERESTS AND EMERGENCE

1. *See* Kurt Gödel, *On Formally Undecidable Propositions of* Principia Mathematica *and Related Systems*, in JEAN VAN HEIJENOORT, FROM FREGE TO GÖDEL: A SOURCE BOOK ON MATHEMATICAL LOGIC 178 (1967); *see also* JOHN D. BARROW, IMPOSSIBILITY: THE LIMITS OF SCIENCE AND THE SCIENCE OF LIMITS 218–47 (1998); PAUL W. GLIMCHER, DECISIONS, UNCERTAINTY, AND THE BRAIN: THE SCIENCE OF NEUROECONOMICS 72 (2003); Giuseppe Dari-Mattiacci, *Gödel, Kaplow, Shavell: Consistency and Completeness in Social Decisionmaking*, 79 CHI.- KENT L. REV. 497 (2004).

2. *Cf.* CLIVE L. SPASH, GREENHOUSE ECONOMICS: VALUES AND ETHICS 267 (2002) ("Optimality . . . is in fact consistency analysis and the best outcome is not guaranteed by the model, but only the choice that is consistent with the assumptions.").

3. BERNARD WILLIAMS, MORALITY: AN INTRODUCTION TO ETHICS 96 (1972).

4. M. J. Farrell, *The Convexity Assumptions in the Theory of Competitive Markets*, 67 J. POL. ECON. 377 (1959).

5. AMARTYA SEN, COLLECTIVE CHOICE AND SOCIAL WELFARE 22 (1970).

6. Guido Calabresi, *The Pointlessness of Pareto: Carrying Coase Further*, 100 Yale L.J. 1211 (1991).

7. *See, e.g.*, Richard A. Posner, *Pragmatic Liberalism versus Classical Liberalism*, 71 U. CHI. L. REV. 659, 666 (2004) ("The Kaldor-Hicks test, also known as wealth maximization, requires only that the winners' gains exceed the losers' losses; compensation of the losers is not required.").

8. *See, e.g.,* Matthew Adler, *Cost-Benefit Analysis, Static Efficiency, and the Goals of Environmental Law,* 31 BOSTON COLLEGE ENVTL. AFF. L. REV. 591 (2004) (providing an account of cost-benefit analysis that seeks to maximize overall well-being according to an objective list account of welfare, rather than a preference-based account); Martha C. Nussbaum, *The Costs of Tragedy: Some Moral Limits of Cost-Benefit Analysis,* 29 J. LEGAL STUD. 1005, 1029–30 (2000) (describing ways in which cost-benefit analysis need not entail utilitarianism or even consequentialism); Amartya Sen, *The Discipline of Cost-Benefit Analysis,* 29 J. LEGAL STUD. 931, 936 (2000) (arguing that cost-benefit analysis can be conceived to allow for "[b]roadly consequential evaluation," including "not only such things as happiness or the fulfillment of desire on which utilitarians tend to concentrate, but also whether certain actions have been performed or particular rights have been violated").

9. Admittedly, academic proponents of cost-benefit analysis often deny that they intend the results of the cost-benefit-analysis calculation to impose a strict pass-fail test in this manner. Proponents outside of the academy, however, often are not as circumspect. For instance, the infamous "supermandate" from House Bill 1022, the Risk Assessment and Cost-Benefit Act of 1995, would have required that cost-benefit-decision criteria supplant any and all conflicting criteria in federal regulatory programs. *See* Thomas O. McGarity, *The Goals of Environmental Legislation,* 31 BOSTON COLLEGE ENVTL. AFF. L. REV. 529, 551 (2004) (describing the bill).

10. Thomas O. McGarity, *A Cost-Benefit State,* 50 ADMIN. L. REV. 7, 39 (1998)

11. *See* STEPHEN BREYER, BREAKING THE VICIOUS CIRCLE: TOWARD EFFECTIVE RISK REGULATION 19–21 (1993).

12. *Cf.* Laurence H. Tribe, *Policy Science: Analysis or Ideology?* 2 PHIL. & PUB. AFF. 66, 85 (1972) (describing use of "the device of an imagined 'impartial spectator'" by proponents of cost-benefit analysis to bolster the appearance of objectivity to their policy analyses).

13. Cass R. Sunstein, "Cost-Benefit Analysis and the Environment" 24 (Univ. of Chicago Law Sch. John M. Olin Program in Law & Econ., Working Paper No. 227, 2d series, 2004), *available at* http://www.law.uchicago.edu/lawecon/index.html.

14. *See, e.g.,* W. Kip Viscusi, *Risk Equity,* 29 J. LEGAL STUD. 843, 845 (2000) ("The functioning of efficient markets involving risk establishes what I will take as my reference point for equitable risks.").

15. *See* Michael Abramowicz, *Toward a Jurisprudence of Cost-Benefit Analysis,* 100 MICH. L. REV. 1708, 1718 (2002); *see also* Herbert Hovenkamp, *Knowledge About Welfare: Legal Realism and the Separation of Law and Economics,* 84 MINN. L. REV. 805, 846 (2000) (providing a historical account of, inter alia, the Progressive Era, pragmatism, and neoclassical economic theory and noting that "the Progressive conception of welfare was not defined by subjectively asserted preference").

16. U.S. Fish & Wildlife Serv., Draft Economic Analysis of Critical Habitat Designation for the Rio Grande Silvery Minnow, Final Draft (May 2002), *cited in* Amy Sinden, *The Economics of Endangered Species: Why Less is More in the Economic Analysis of Critical Habitat Designations*, 28 Harv. Envtl. L. Rev. 129, 169–70 (2004).

17. *See* Amartya Sen, Commodities and Capabilities (1985).

18. *See* Andrew Lang, *The GATS and Regulatory Autonomy: A Case Study of Social Regulation of the Water Industry*, 7 J. Int'l Econ. L. 801, 808 (2004) (describing the common use of "direct subsidy payments to consumers" in order to address access problems following water-market reform).

19. *See* Louis Kaplow & Steven Shavell, Fairness versus Welfare 33 (2002).

20. Mary Wollstonecraft, A Vindication of the Rights of Woman 167 (Penguin Books, 1992) (1792). Wollstonecraft may have been inspired by Kant, who posited that the duty of hospitality toward members of the world community was "not a question of philanthropy but of right." *See* Seyla Benhabib, Another Cosmopolitanism 22 (2006) (quoting Kant's 1795 work *Perpetual Peace: A Philosophical Sketch*).

21. *See* Reed D. Benson, *Recommendations for an Environmentally Sound Policy on Western Water*, 17 Stan. Envtl. L.J. 247, 263 (1998) (noting that tiered pricing "is a key tool" for water management).

22. *See* W. Adamowicz et al., *Combining Revealed and Stated Preference Methods for Valuing Environmental Amenities*, 26 J. Envtl. Econ. & Mgmt. 271 (1994).

23. *See* Kelly B. Macguire et al., "Willingness to Pay to Reduce a Child's Pesticide Exposure: Evidence from the Baby Food Market" (Nat'l Ctr. For Envtl. Econ., Working Paper No. 02-03, 2002), *available at* http://yosemite.epa.gov/EE/epa/eed.nsf/WPNumberNew/2002-03?OpenDocument.

24. Viscusi, *supra* note 14, at 849 (emphasis added).

25. *See* Mark Sagoff, Price, Principle, and the Environment 77 (2004).

26. *See* Spash, *supra* note 2, at 170–71 (describing William D. Nordhaus, "New Estimates of the Economic Impacts of Climate Change" [unpublished paper, 1998]).

27. Two recent meta-analyses are W. Kip Viscusi & Joseph E. Aldy, *The Value of a Statistical Life: A Critical Review of Market Estimates throughout the World*, 27 J. Risk & Uncertainty 5, 44 (2003); and Ikuho Kochi, Bryan Hubbell & Randall Kramer, *An Empirical Bayes Approach to Combining and Comparing Estimates of the Value of a Statistical Life for Environmental Policy Analysis*, 34 Envtl. & Resource Econ. 385, 400 (2006).

28. *See* Elizabeth Anderson, Value in Ethics and Economics 105–203 (1993); Sidney A. Shapiro & Robert L. Glicksman, Risk Regulation at Risk: Restoring a Pragmatic Approach 98–100 (2003).

29. *See* Mark Sagoff, The Economy of the Earth: Philosophy, Law, and the Environment 116 (1988) ("A hundred years of compassionate legislation has produced conditions in which economists now argue that voluntary

markets set an appropriate value on worker safety. This is a result not of more efficient markets but of persistent ethical regulation.").

30. *Cf.* SHAPIRO & GLICKSMAN, *supra* note 28, at 100 (noting that "[t]he pool of labor for hazardous jobs . . . consists of disadvantaged workers who are willing to accept health and safety risks in return for very modest amounts of compensation") (internal quotation marks omitted).

31. *See* Kochi, Hubbell & Kramer, *supra* note 27, at 399 (finding within a sample of U.S. studies a mean value of $17.0 million for union workers and only $6.8 million for nonunion); Viscusi & Aldy, *supra* note 27, at 44. ("Regardless of the estimation strategy, most assessments of the U.S. labor market found higher risk premiums for union workers than for non-union workers.")

32. Eric A. Posner & Cass R. Sunstein, *Dollars and Death,* 72 U. CHI. L. REV. 537, 538 (2005).

33. *See* Sunstein, "Cost-Benefit Analysis and the Environment," *supra* note 13, at 25 ("In many cases of environmental regulation . . . rights violations are not involved; we are speaking here of statistically small risks.").

34. *Id.* at 24. For a summary of the psychological literature, see Douglas A. Kysar, *The Expectations of Consumers,* 103 COLUM. L. REV. 1700, 1763–66 (2003).

35. Lisa Heinzerling, *The Rights of Statistical People,* 24 HARV. ENVTL. L. REV. 189 (2000); *see also* Frank I. Michelman, *Pollution as a Tort: A Non-Accidental Perspective on Calabresi's Costs,* 80 YALE L.J. 547 (1971).

36. *See* ANDERSON, *supra* note 28, at 144–47, 158–59, 203–10; SAGOFF, *supra* note 29, at 7–14; CASS R. SUNSTEIN, FREE MARKETS AND SOCIAL JUSTICE 21–23, 44–45, (1997).

37. SAGOFF, *supra* note 29, at 70.

38. *See* Cass R. Sunstein, *Valuing Life: A Plea for Disaggregation,* 54 DUKE L.J. 385 (2004).

39. *Id.* at 391.

40. *See id.* at 405 ("If wealthy people show a higher [willingness-to-pay] than poor people, then a uniform [willingness-to-pay] based on a population-wide median will ensure insufficient protection of wealthy people and excessive protection of poor people in a way that might well prove harmful to both groups.").

41. *See* Kimberly A. Yuracko, *Private Nurses and Playboy Bunnies: Explaining Permissible Sex Discrimination,* 92 CAL. L. REV. 147 (2004).

42. Office of Mgmt. and Budget, Draft 2003 Report to Congress on the Costs and Benefits of Federal Regulations, 68 Fed. Reg. 5492, 5499 (Feb. 3, 2003).

43. W. Kip Viscusi & Richard J. Zeckhauser, *Sacrificing Civil Liberties to Reduce Terrorism Risks,* 26 J. RISK & UNCERTAINTY 99, 104–05 (2003).

44. *Id.* at 105, tbl. 1.

45. *See* CASS R. SUNSTEIN, FREE MARKET AND SOCIAL JUSTICE 17 (1997) ("[W]hen preferences are a function of legal rules, the government cannot take preferences as given [and] the rules cannot be justified by reference to the preferences. . . ."); Cass R. Sunstein, *Endogenous Preferences, Environmental Law,*

22 J. LEGAL STUD. 217 (1993). *Cf.* Samuel Bowles, *Endogenous Preferences: The Cultural Consequences of Markets and Other Economic Institutions,* 36 J. ECON. LIT. 75, 75 (1998) (describing conceptual problems created for economic theory when markets "influence the evolution of values, tastes, and personalities").

46. LOUIS MENAND, THE METAPHYSICAL CLUB 299 (2001) (quoting letter from John Dewey).

47. As Laurence Tribe observes, "the whole point of personal or social choice in many situations is not to implement a given system of values in light of the perceived facts, but rather to define, and sometimes deliberately to reshape, the values—and hence the identity—of the individual or community that is engaged in the process of choosing." Laurence H. Tribe, *Policy Science: Analysis or Ideology?* 2 PHIL. & PUB. AFF. 66, 99 (1972).

48. Sunstein, "Cost-Benefit Analysis and the Environment," *supra* note 13, at 24.

49. *Cf.* Harry G. Frankfurt, *Freedom of the Will and the Concept of a Person,* 68 J. PHIL. 5 (1971) ("The statement that a person enjoys freedom of the will means that he is free to want what he wants to want."). As Henry Richardson has noted, often the task of determining what one wants to want entails dialogue and interaction with others. *See* Henry S. Richardson, *The Stupidity of the Cost-Benefit Standard,* in COST-BENEFIT ANALYSIS: LEGAL, ECONOMIC, AND PHILOSOPHICAL PERSPECTIVES 135, 158 (Matthew D. Adler & Eric A. Posner eds., 2001) ("[I]ndividual' choices are, across a very wide range, worked out in dynamic compromise with the preferences of the many others with whom they interact.").

CHAPTER 5. OTHER STATES

1. REPORT OF THE UNITED NATIONS CONFERENCE ON ENVIRONMENT AND DEVELOPMENT, Annex I, at Principle 2, U.N. Doc. A/CONF.151/26 (1992) (*Rio Declaration on Environment and Development*).

2. *Id.*

3. 22 U.S.C.A. § 287 note (2009).

4. 22 U.S.C.A. § 274a (2009).

5. 22 U.S.C.A. § 2151p (2009).

6. *Id.*

7. 33 U.S.C.A. § 1251(c) (2009); 33 U.S.C.A. § 1419 (2009).

8. *See, e.g., Wilderness Society v. Morton,* 463 F.2d 1261 (D.C. Cir. 1972); *Province of Manitoba v. Norton,* 398 F. Supp. 2d 41 (D.D.C. 2005).

9. *See* Sanford E. Gaines, *"Environmental Effects Abroad of Major Federal Actions": An Executive Order Ordains a National Policy,* 3 HARV. ENVTL. L. REV. 136 (1979). Although the United States is not a party to the agreement, the "Convention on Environmental Impact Assessment in a Transboundary Context," which entered into force in 1997, provides extensive extraterritorial impact-assessment procedures for member parties. "Convention on Environmental Impact in a Transboundary Context," Feb. 25, 1991, 30 I.L.M.

800 (1991), *available at* http://www.unece.org/env/eia/documents/conven
tiontextenglish.pdf.

10. *Cf.* Judith Resnick & Julie Chi-hye Suk, *Adding Insult to Injury: Questioning
the Role of Dignity in Conceptions of Sovereignty*, 55 STAN. L. REV. 1921, 1928
(2003) (suggesting that the recognition of role-dignitary interests of states
does not necessarily cut in favor of sovereign immunity, but can instead be
read to require sovereigns to explain their conduct).

11. *See* BRUCE A. ACKERMAN, SOCIAL JUSTICE IN THE LIBERAL STATE 100–103
(1980).

12. *See* JÜRGEN HABERMAS, BETWEEN NATURALISM AND RELIGION 320 (Ciaran
Cronin trans., 2008) ("If even a superpower can no longer guarantee the security and welfare of its own population without the help of other states, then
'sovereignty' is losing its classical meaning.").

13. *See* WORLD BANK, THE WORLD DEVELOPMENT REPORT (1992).

14. Theodore Panayotou, "Economic Growth and the Environment" (Ctr. for Int'l
Dev., Harv. Univ., Working Paper No. 56, 2000), *available at* http://www.cid
.harvard.edu/cidwp/056.htm.

15. *See* BJØRN LOMBORG, THE SKEPTICAL ENVIRONMENTALIST: MEASURING THE
REAL STATE OF THE WORLD 210 (2001) ("It is . . . reasonable to expect that as
the developing countries of the world achieve higher levels of income, they
will—as we in the developed world have done—opt for and be able to afford
an ever clearer environment."). A more recent and slightly more refined version of this claim can be found in TED NORDHAUS & MICHAEL SHELLEN-
BERGER, BREAK THROUGH: FROM THE DEATH OF ENVIRONMENTALISM TO THE
POLITICS OF POSSIBILITY (2007).

16. See Simone Borghesi, "The Environmental Kuznets Curve: A Survey of the Literature" 4 (1999) (municipal solid waste), available athttp://www.feem.it/NR/
rdonlyres/1D089671-FFCF-42F9-BA15-DEB9E2A581F1/138/8599.pdf; S. M.
De Bruyn et al., *Economic Growth and Emissions: Reconsidering the Empirical Basis of Environmental Kuznets Curves*, 25 ECOL. ECON. 161 (1998) (nitrogen ox-
ides); Susmita Dasgupta et al., *Confronting the Environmental Kuznets Curve*,
16 J. ECON. PERSP. 147, 162–63 (2002) (toxics); Daniel C. Esty, *Bridging the
Trade-Environment Divide*, 15 J. ECON. PERSP. 113, 115 (2001) (greenhouse gases).

17. *See* Hemamala Hettige et al., "Industrial Pollution in Economic Development: Kuznets Revisisted" (World Bank Pol'y Res. Working Paper No. 1876,
Jan. 31, 1999) (water pollution), *available at* http://www-wds.worldbank.org/
servlet/WDS_IBank_Servlet?pcont=details&eid=000009265
_3980312102605.

18. COREY L. LOFDAHL, ENVIRONMENTAL IMPACTS OF GLOBALIZATION AND TRADE:
A SYSTEMS STUDY 121–22 (2002).

19. LOMBORG, *supra* note 15, at 117.

20. *See Rio Declaration*, *supra* note 1, at Principle 12 ("States should cooperate to
promote a supportive and open international economic system that would
lead to economic growth and sustainable development in all countries. . . .").

21. DAVID RICARDO, THE PRINCIPLES OF POLITICAL ECONOMY AND TAXATION 82–87 (Michael P. Fogarty ed., J. M. Dent & Sons 1962) (1817).

22. *See id.* at 83. The relative lack of attention devoted to the assumption of international capital immobility in trade debates has driven some commentators to engage in rather unusual efforts to underscore its importance to Ricardo's thinking. *See* Roy J. Ruffin, *David Ricardo's Discovery of Comparative Advantage*, 34 HIST. POL. ECON. 727, 734 (2002) (noting that "of the 973 words Ricardo devoted to explaining the law of comparative advantage, 485 emphasized the importance of factor immobility").

23. Paul A. Samuelson, *Where Ricardo and Mill Rebut and Confirm Arguments of Mainstream Economists Supporting Globalization*, 18 J. ECON. PERSP. 135, 143 (2004).

24. For instance, in developing his "invisible hand" passage, Adam Smith was careful to point out that the capitalist tends to "prefer[] the support of domestic to that of foreign industry," even if more generally he is driven by "only his own gain." ADAM SMITH, WEALTH OF NATIONS 423 (Edwin Cannan ed., Random House 1937) (1776). Similarly, Ricardo noted with approval that feelings of nationalist loyalty cause "most men of property to be satisfied with a low rate of profits in their own country, rather than seek a more advantageous employment for their wealth in foreign nations." RICARDO, *supra* note 21, at 83. More recently, John Maynard Keynes emphasized that the "divorce between ownership and the real responsibility of management" becomes "intolerable" when "applied internationally." John Maynard Keynes, *National Self-Sufficiency*, NEW STATESMAN & NATION, July 8 & 15, 1933, reprinted in THE COLLECTED WRITINGS OF JOHN MAYNARD KEYNES 233, 236 (Donald Moggridge ed., 1982).

25. *See, e.g.,* DAVID VOGEL, TRADING UP: CONSUMER AND ENVIRONMENTAL REGULATION IN A GLOBAL ECONOMY (1995).

26. *Rio Declaration, supra* note 1, at Principle 22.

27. For discussion of alternative conceptions of consumer decision making and the impact of advertising and other marketing practices, see Douglas A. Kysar, *The Expectations of Consumers*, 103 COLUM. L. REV. 1700 (2003); Douglas A. Kysar, *Kids & Cul-de-Sacs: Census 2000 and the Reproduction of Consumer Culture*, 87 CORNELL L. REV. 853 (2002).

28. HERMAN DALY, BEYOND GROWTH 149 (1997).

29. PHILIP ALLOTT, THE HEALTH OF NATIONS 407 (2002).

30. William Nordhaus exemplifies this approach in his dismissal of *The Stern Review*'s attempt to prescribe a morally appropriate discount rate in the context of global climate-change-policy analysis: "The normatively acceptable real interest rates prescribed by philosophers, economists, or the British government are irrelevant to determining the appropriate discount rate to use in the actual financial and capital markets of the United States, China, Brazil, and the rest of the world. When countries weigh their self-interest in international bargains about emissions reductions and burden sharing, they will

look at the actual gains from bargains, and the returns on these relative to other investments, rather than the gains that would come from a theoretical growth model." William D. Nordhaus, *A Review of* The Stern Review on the Economics of Climate Change, 45 J. ECON. LIT. 686, 692 (2007).

31. PAUL W. KAHN, PUTTING LIBERALISM IN ITS PLACE 45 (2005).

32. *See, e.g.,* ACKERMAN, *supra* note 11, at 95 (concluding that "[t]he *only* reason for restricting immigration is to protect the ongoing process of liberal conversation itself") (emphasis in original).

33. *See, e.g.,* THOMAS W. POGGE, WORLD POVERTY AND HUMAN RIGHTS: COSMO-POLITAN RESPONSIBILITIES AND REFORMS (2002); THOMAS W. POGGE, REALIZING RAWLS (1989).

34. Pavlos Eleftheriadis, *The Idea of a European Constitution,* 27 OXFORD J. LEGAL STUD. 1, 19 (2007). For a more nuanced view, but one that still expresses significant skepticism regarding the role of justice theory beyond national borders, see Thomas Nagel, *The Problem of Global Justice,* 33 PHIL. & PUB. AFF. 113 (2005). For trenchant criticisms of Nagel's view, see Joshua Cohen & Charles Sabel, *Extra Rempublicam Nulla Justitia?* 34 PHIL. & PUB. AFF. 147 (2006); and A. J. Julius, *Nagel's Atlas,* 34 PHIL. & PUB. AFFAIRS 176 (2006). For an argument that "[o]nly the transformed consciousness of citizens, as it imposes itself in areas of domestic policy, can ultimately pressure global actors to change their own self-understanding sufficiently to begin to see themselves as members of an international community who are compelled to cooperate with one another, and hence to take another's interests into account," see JÜRGEN HABERMAS, THE POSTNATIONAL CONSTELLATION: POLITICAL ESSAYS 55 (Max Pensky ed., 2001). And for what is perhaps the most defensible claim of all in this area, see WILL KYMLICKA, CONTEMPORARY POLITICAL PHILOSOPHY: AN INTRODUCTION 313 (2d. ed. 2002). ("[F]ew people have any clear idea what principles of justice or standards of democratization or norms of virtue or loyalty should apply to transnational institutions.")

35. JOHN RAWLS, THE LAW OF PEOPLES 120 (1999).

36. *See* Wojciech Kopczuk, Joel Slemrod & Shlomo Yitzhaki, *The Limitations of Decentralized World Redistribution: An Optimal Taxation Approach,* 49 EUROPEAN ECON. REV. 1051, 1072 (2005).

37. Jack Goldsmith & Eric A. Posner, *The New International Law Scholarship,* 34 GA. J. INT'L & COMP. L. 463, 472 (2006). *See also* JACK L. GOLDSMITH & ERIC A. POSNER, THE LIMITS OF INTERNATIONAL LAW (2005).

38. Eric A. Posner, *International Law: A Welfarist Approach,* 73 U. CHI. L. REV. 487, 501 (2006).

39. *Id.* at 504.

40. PHILIP ALLOTT, INTERNATIONAL LAW AND INTERNATIONAL REVOLUTION: RECONCEIVING THE WORLD 16 (1989).

41. *See also* Ryan Goodman & Derek Jinks, *Toward an Institutional Theory of Sovereignty,* 55 STAN. L. REV. 1749, 1786 (2003). The authors note that "several

constitutive features of the modern state (including the very notion of being an autonomous actor) are socially constructed at a global level."

42. Goldsmith & Posner, *supra* note 37, at 463.

43. *See* Alexander Wendt, *The State as a Person in International Theory*, 30 Rev. Int'l Stud. 289, 316 (2004) (noting that even the most empirically inclined international relations scholars "are not objective observers of a separate reality, but part of that reality, and as such are at least indirectly responsible for its effects").

44. Goldsmith & Posner, *supra* note 37, at 463.

45. *Id.* at 466. *Cf.* Nagel, *supra* note 34, at 140 ("Justice applies, in other words, only to a form of organization that claims political legitimacy and the right to impose decision by force, and not to a voluntary association or contract among independent parties concerned to advance their common interests.").

46. *Cf.* Robert Nozick, Anarchy, State, and Utopia 110 (1974) (observing that "[u]tilitarian theory is embarrassed by the possibility of utility monsters who get enormously greater sums of utility from any sacrifice of others than these others lose").

47. Goldsmith & Posner, *supra* note 37, at 469 (arguing that "compliance with a treaty should decline as the number of state parties increases").

48. *Id.* at 468 (emphasis added).

49. 42 U.S.C.A. § 7521(a)(1) (2009).

50. 549 U.S. 497 (2007).

51. Transcript of Oral Argument, *Massachusetts v. EPA*, 549 U.S. 497 (2007), No. 05-1120, at 50, *available at* www.supremecourtus.gov/oral_arguments/argument_transcripts/05-1120.pdf.

52. *See* Jonathan Glover, *"It Makes No Difference Whether or Not I Do It,"* 49 Proceedings Aristotelian Soc'y (Suppl.) 171 (1975).

53. 127 S. Ct. at 1451. For an extensive analysis of this argument in connection with the constitutional permissibility of state climate change regulation, see Douglas A. Kysar & Bernadette A. Meyler, *Like a Nation State*, 55 UCLA L. Rev. 1621 (2008)

54. Transcript of Oral Argument, *supra* note 51, at 12.

55. *Id.* at 50.

56. *Id.* at 55.

57. Christine Todd Whitman, *Carbon Ruling: A Welcome First Step*, Washington Post, Apr. 9, 2007.

58. John Vidal, *U.S. Rejects All Proposals on Climate Change*, Guardian, May 26, 2007, *available at* http://www.guardian.co.uk/world/2007/may/26/usa.greenpolitics.

59. Darren Samuelsohn, *House Transportation Chair, Bush Admin. Oppose E.U. Airline Plan*, Greenwire, May 11, 2007.

60. "President Bush Meets with EU Leaders, Chancellor Merkel of the Federal Republic of Germany and President Barroso of the European Council and

President of the European Commission," transcript *available at* http://merln
.ndu.edu/archivepdf/EUR/WH/20070430-2.pdf.

61. Andrew Grice, *Global Warming: The Climate Has Changed*, THE INDEPEN-
DENT, Mar. 14, 2007, *available at* http://www.independent.co.uk/environment/
climate-change/global-warming-the-climate-has-changed-440117.html.

62. *Id.*

63. Shingo Ito, *On Eve of G8 Summit, E.U. Reaches Agreement with Japan on Emissions Cuts*, AGENCE FRANCE-PRESSE, June 5, 2007.

64. Likewise, reports suggest that Danish officials and citizens felt significant responsibility for the success of negotiations for a post-Kyoto climate agreement, in light of the key talks being held in Copenhagen. *See* Lisa Friedman,
2009 Climate Talks Are Already a Nail Biter for the Danes, GREENWIRE, July 14,
2008 (quoting Angela Anderson, director of the Pew Environment Group's
global warming campaign: "They want their name on this agreement. They
want it to go down in history as the moment the world really got on top of
this problem.").

65. *See, e.g.*, Scott Barrett, *The Problem of Averting Global Catastrophe*, 6 CHI. J.
INT'L L. 527 (2006); Bruce Yandle & Stuart Buck, *Bootleggers, Baptists, and the
Global Warming Battle*, 26 HARV. ENVTL. L. REV. 177 (2002).

66. *See* Transcript of Oral Argument, *supra* note 51, at 9 (referring to the challengers' affidavits: "Those affidavits talked about the fact that if the government
starts to regulate, the technology is going to change, if the technology
changes, other governments will adopt it, and all, and that strikes me as sort
of spitting out conjecture on conjecture. . . ."); *see also Massachusetts v. EPA*,
127 S. Ct. at 1469 (Roberts, C.J., dissenting) (characterizing the argument as
"do not worry that other countries will contribute far more to global warming
than will U.S. automobile emissions; someone is bound to invent something,
and places like the People's Republic of China or India will surely require use
of the new technology, regardless of cost").

67. 127 S. Ct. at 1469.

68. *ASARCO, Inc. v. Kadish*, 490 U.S. 605, 615 (1989) (Kennedy, J.) (arguing that
where an element of standing "depends on the unfettered choices made by
independent actors not before the courts and whose exercise of broad and legitimate discretion the courts cannot presume either to control or predict,"
the party seeking standing must present evidence in support of their claim as
to how the independent actors will behave).

69. 127 S. Ct. at 1471.

70. *Id.* at 1457.

71. Leading legal scholars similarly dismiss corrective justice arguments in favor
of disproportionate U.S. responsibility for greenhouse gas mitigation on the
theory that prior emissions restrictions would have had no discernible effect
on the overall climate. *See* Eric A. Posner & Cass R. Sunstein, *Climate Change
Justice*, 96 GEO. L.J. 1565, 1600 (2008).

72. The most ambitious of these international agreements, the 1979 "Geneva Convention on Long-Range Transboundary Air Pollution," obligates its fifty-one member parties, including the United States, to "endeavour to limit and, as far as possible, gradually reduce and prevent air pollution including long-range transboundary air pollution." "Convention on Long-Range Transboundary Air Pollution," Nov. 13, 1979, T.I.A.S. No. 10541, reprinted in 18 I.L.M. 1442. Subsequent protocols to the convention have spelled out more specific air pollution reduction and prevention obligations but have attracted fewer ratifications.

73. James J. Yienger et al., *The Episodic Nature of Air Pollution Transport from Asia to North America*, 108 J. Geophysical Res. 26931, 26944 (2000).

74. A summary table of all current national ambient air quality standards is maintained at http://epa.gov/air/criteria.html.

75. For this work, *background ozone* means "air from the Pacific with O_3 levels not significantly influenced by North American emissions within the previous three days." Daniel Jaffe et al., *Increasing Background Ozone During Spring on the West Coast of North America*, 30 Geophysical Res. Letters 15-1 (2003).

76. Robert E. Lutz, *Managing a Boundless Resource: U.S. Approaches to Transboundary Air Quality Control*, 11 Envtl. L. 321 (1981).

77. 127 S. Ct. at 1454.

78. Brian Barry, for instance, has observed that Hobbesian ruthlessness seems to be exactly what past generations have shown toward the present, leading him to doubt prospects for a dramatic reorientation of attitudes toward futurity. *See* Brian Barry, Democracy, Power, and Justice: Essays in Political Theory 485 (1989) ("My impression is that the only reason why our ancestors did not do more damage is that they lacked the technology to do so.").

79. W. Kip Viscusi, *Rational Discounting for Regulatory Analysis*, 74 U. Chi. L. Rev. 209, 211 (2007).

80. *See generally* John O'Neil, Ecology, Policy, and Politics: Human Well-Being and the Natural World (1993).

CHAPTER 6. OTHER GENERATIONS

1. For an early and insightful articulation of this viewpoint, see *Philippe J. Sands, The Environment, Community and International Law*, 30 Harv. Int'l L.J. 393 (1989).

2. *See* Thomas Nagel, *The Problem of Global Justice*, 33 Phil. & Pub. Aff. 113, 113 (2005) ("However imperfectly, the nation-state is the primary locus of political legitimacy and the pursuit of justice, and it is one of the advantages of domestic political theory that nation-states actually exist.").

3. Although focus in the text remains on the idea of nation-state subjectivity, as Will Kymlicka has observed, "the problem of social unity and political stability" might also be addressed by "an emphasis on a common way of life" or "an emphasis on political participation." Will Kymlicka, Contemporary Political Philosophy: An Introduction 257 (2d ed. 2002). To some

extent, the precautionary approach can be thought to marry the ideas of common nationhood and political participation by structuring a broadly inclusive policy dialogue regarding the content of the state's ethical relations toward environmental others. It is even possible to imagine a common way of life emerging from such a marriage, as the state's environmental identity comes to entail rather dramatic alterations to consumption, transportation, housing, employment, and other patterns that determine impact. For further discussion of the concept of a common nationality, see David Miller, *Community and Citizenship*, in COMMUNITARIANISM AND INDIVIDUALISM 85 (Shlomo Avineri & Avner De-Shalit eds., 1992).

4. Steven Shiffrin, *Government Speech*, 27 UCLA L. REV. 565, 647 (1980).

5. *Cf.* Richard P. Hiskes, *The Right to a Green Future: Human Rights, Environmentalism, and Intergenerational Justice*, 27 HUM. R'TS Q. 1346, 1350 (2005) (noting that future generations' "countenances seem ineluctably lost as faces in a very abstract crowd—as members of a group that we can imagine but to which we have a hard time extending group human rights").

6. 42 U.S.C.A. § 4331(b)(1) (2009); *see also* National Park Service Organic Act, 16 U.S.C.A. § 1 (2009) (declaring that the purpose of national parks, monuments, and reservations is "to conserve the scenery and the natural and historic objects and the wild life therein, and to provide for the enjoyment of the same in such manner and by such means as will leave them unimpaired for the enjoyment of future generations"); Wilderness Act of 1964, 16 U.S.C.A. § 1131(a) (2009) (announcing "the policy of the Congress to secure for the American people of present and future generations the benefits of an enduring resources of wilderness").

7. IROQUOIS NATIONS CONSTITUTION, Tree of the Long Leaves § LI, reprinted in ARTHUR C. PARKER, THE CONSTITUTION OF THE FIVE NATIONS 38–39 (Iroqrafts Ltd. 1991) (1916).

8. *See, e.g.*, WILFRED BECKERMAN, A POVERTY OF REASON: SUSTAINABLE DEVELOPMENT AND ECONOMIC GROWTH (2002); W. Kip Viscusi, *Rational Discounting for Regulatory Analysis*, 74 U. CHI. L. REV. 209 (2007).

9. THOMAS PAINE, THE RIGHTS OF MAN 36 (Everyman's Library ed. 1958) (1791); *see also* EDMUND BURKE, REFLECTIONS ON THE REVOLUTION IN FRANCE 192 (Conor Cruise O'Brien ed., 1968) (1790) ("[O]ne of the first and most leading principles on which the commonwealth and the laws are consecrated, is lest the temporary possessors and life-renters in it, unmindful of what they have received from their ancestors, or what is due to their posterity, should act as if they were the entire masters; that they should not think it amongst their rights to cut off the entail, or commit waste on the inheritance, by destroying at their pleasure the whole fabric of their society. . . .").

10. THEODORE ROOSEVELT, A BOOK-LOVER'S HOLIDAYS IN THE OPEN (1916), *available at* http://www.bartleby.com/57/.

11. *Id.*

12. JOHN RAWLS, A THEORY OF JUSTICE 285, 288 (1971).

13. JOHN RAWLS, JUSTICE AS FAIRNESS: A RESTATEMENT 159 (2001).

14. RAWLS, *supra* note 12, at 289; *see also* JOHN RAWLS, POLITICAL LIBERALISM 245–46 (1993) (suggesting that society can preserve nature and biodiversity because of its instrumental and aesthetic worth to individuals, but that to adopt an "attitude of natural religion" would violate the norms of public reason).

15. *See* Ronald Dworkin, *Liberalism,* in LIBERALISM AND ITS CRITICS 60, 76–77 (Michael J. Sandel ed., 1984).

16. BRUCE A. ACKERMAN, SOCIAL JUSTICE IN THE LIBERAL STATE 217 (1980).

17. *See id.* at 212 (stating that "over a large area of the economy, it will make sense to ask whether the [present] generation has, through capital formation and technological innovation, passed on a stock of manna-like resources that permits [future generations] to begin in a material position made no worse by the passage of time").

18. *But see* Jeremy Waldron, *Enough and as Good Left for Others,* 29 PHIL. Q. 319, 319–28 (1979) (offering a powerful argument that Locke intended the "enough, and as good" proviso to establish a sufficient, but not a necessary, condition for legitimate appropriation of resources).

19. Rawls at one point supposed that "there is no society anywhere in the world—except for marginal cases—with resources so scarce that it could not, were it reasonably and rationally organized and governed, become well-ordered." RAWLS, *supra* note 12, at 108 n.34.

20. *See* Ocean World, "Coral Reefs: Coral Reef Destruction and Conservation," http://oceanworld.tamu.edu/students/coral/coral5.htm.

21. *See* ACKERMAN, *supra* note 16, at 212–17 (discussing a similarly intractable conflict between generations and concluding that the process of liberal education, whereby citizens are taught to respect the plurality of value, offers the best hope of lessening conflict by encouraging present generations to reconsider their goals in light of future generations' goals).

22. UN DEP'T OF ECON. & SOC. AFF., DIV. OF SUSTAINABLE DEV., PLAN OF IMPLEMENTATION OF THE WORLD SUMMIT ON SUSTAINABLE DEVELOPMENT 2 (2002), *available at* http://www.un.org/esa/sustdev/documents/WSSD_POI_PD/English/WSSD_PlanImpl.pdf.

23. They also benefited from libertarian schools of thought, which argued that the problem of intergenerational justice is best resolved through decentralized, private-market activity: "A classical liberal regime of limited government, personal liberty, and private property benefits future generations more than an alternative regime that consciously enlists large government to restrain liberty and to limit the present use of property for the benefit of future generations." Richard A. Epstein, *Justice Across the Generations,* 67 TEX. L. REV. 1465, 1466 (1989).

24. *See, e.g.,* ACKERMAN, *supra* note 16, at 202 (observing that the "problem of inheritance is of such great theoretical importance that we must confront it head-on if we hope to grasp the shape of liberal ideals"); DEREK PARFIT,

REASONS AND PERSONS 351 (1984) (arguing that development norms of inter-generational responsibility "is the most important part of our moral theory, since the next few centuries will be the most important in human history"); RAWLS, *supra* note 12, at 284 (observing that "the question of justice between generations . . . subjects any ethical theory to severe if not impossible tests"); Amartya K. Sen, *On Optimising the Rate of Savings*, 71 ECON. J. 479, 486 (1961) (observing that there can be no democratic solution to intergenera-tional problems); Lawrence B. Solum, *To Our Children's Children's Children: The Problems of Intergenerational Ethics*, 35 LOY. L.A. L. REV. 163, 164 (2001) (stating that "[t]he problems of interegenerational ethics are notoriously some of the most difficult in moral and political philosophy").

25. *See* PARTHA S. DASGUPTA & GEOFFREY M. HEAL, ECONOMIC THEORY AND EXHAUSTIBLE RESOURCES 472 (1979).

26. *Id.*

27. *See* Kenneth Arrow et al., *Are We Consuming Too Much?* 18 J. ECON. PERSP., Summer 2004, 147, 154–55.

28. *See id.* at 155.

29. *See* Robert N. Stavins, Alexander F. Wagner & Gernot Wagner, *Interpreting Sustainability in Economic Terms: Dynamic Efficiency Plus Intergenerational Eq-uity*, 79 ECON. LETTERS 339, 342 (2003).

30. *Id.* at 342.

31. *Id.* at 343.

32. Tyler Cowen & Derek Parfit, *Against the Social Discount Rate*, in JUSTICE BE-TWEEN AGE GROUPS AND GENERATIONS 144, 147 (Peter Laslett & James S. Fishkin eds., 1992).

33. Although many conclude that discounting within a single generation is un-problematic, as Lisa Heinzerling has shown, the practice still entails a variety of conceptually and morally challenging questions. *See* Lisa Heinzerling, *Dis-counting Our Future*, 34 LAND & WATER L. REV. 39 (1999); Lisa Heinzerling, *Environmental Law and the Present Future*, 87 GEO. L.J. 2025 (1999).

34. *See* CLIVE L. SPASH, GREENHOUSE ECONOMICS: VALUE AND ETHICS 201–03 (2002). Indeed, one of the main reasons that *The Stern Review on the Econom-ics of Climate Change* report differed so dramatically and controversially from other prominent economic analyses of climate change was that the report's authors departed from economic orthodoxy with respect to intergenerational discounting. *See* William D. Nordhaus, *A Review of* The Stern Review on the Economics of Climate Change, 45 J. ECON. LITERATURE, 686, 701 (2007) ("*The [Stern] Review*'s unambiguous conclusions about the need for extreme immediate action will not survive the substitution of assumptions that are more consistent with today's marketplace real interest rates and savings rates."). Even the impact of uncertainty about the proper rate of discount among economists has been inadequately appreciated. *See* Martin L. Weitzman, *Why the Far-Distant Future Should Be Discounted at Its Lowest Possi-ble Rate*, 36 J. ENVTL. ECON. & MGMT. 201 (1998) (demonstrating that a con-

sensus discount rate, if sought, should be skewed toward the low end of plausible rates in order to offset the impact of compounding interest, which would arbitrarily favor those who support higher rates).

35. *See* FRANK S. ARNOLD, ECONOMIC ANALYSIS OF ENVIRONMENTAL POLICY AND REGULATION 180 (1995) ("It then seems reasonable to discount the future benefits to the present using the same rate that the affected citizens would use, for it is on their behalf that the project is undertaken."); Robert W. Hahn, *The Economic Analysis of Regulation: A Response to the Critics,* 71 U. CHI. L. REV. 1021, 1026 (2004) ("The basic rationale for discounting is that consumers are not indifferent between consuming a dollar's worth of a good today and one dollar next year; discount rates are necessary to reflect this preference.").

36. *See* Lisa Heinzerling, *Regulatory Costs of Mythic Proportions,* 107 YALE L.J. 1981, 2048 (1998).

37. Kenneth J. Arrow et al., *Intertemporal Equity, Discounting, and Economic Efficiency,* in CLIMATE CHANGE 1995: ECONOMIC AND SOCIAL DIMENSIONS OF CLIMATE CHANGE 125, 133 (James P. Bruce, Hoesung Lee & Erik F. Haites eds., 1996).

38. *See* Richard L. Revesz, *Environmental Regulation, Cost-Benefit Analysis, and the Discounting of Human Lives,* 99 COLUM. L. REV. 941, 998 (1999) (describing "[t]he ethically compromised status of discounting for time preference at a constant rate" across, rather than within, generations); Geoffrey Heal, "Climate Economics: A Meta-Review and Some Suggestions" 8 (Nat'l Bureau of Econ. Research, Working Paper No. 13,927, 2008), *available at* http://www.nber.org/papers/w13927 ("In choosing the [rate-of-time preference] the issue is quite simply whether we want to discriminate against future people. I have never seen a convincing explanation for why this is the right thing to do.").

39. Arrow et al., *supra* note 27, at 148.

40. *See* HERMAN E. DALY, BEYOND GROWTH (1996).

41. Along these lines, W. Kip Viscusi observes that future generations are likely to be "more affluent and better off economically than we are," and therefore asserts that "[t]he current citizenry . . . might not be too moved by the plight of their more affluent, distant descendants." Viscusi, *supra* note 8, at 210; *see also* CASS R. SUNSTEIN, WORST-CASE SCENARIOS 12 (2007) ("If human history is any guide, the future will be much richer than the present; and it makes no sense to say that the relatively impoverished present should transfer its resources to the far wealthier future."); Robert C. Lind, *Intergenerational Equity, Discounting, and the Role of Cost-Benefit Analysis,* 23 ENERGY POL'Y 379, 382 (1995) ("[I]n all likelihood future generations will be much richer than the present one and if they (future generations) want lower levels of greenhouse gases and lower temperature levels they should pay for them."). Even John Rawls, who deserves credit for reviving interest in the problem of intergenerational justice, seems to implicitly assume a unidirectional upward trend in

progress and well-being. *See* RAWLS, *supra* note 12, at 253 (suggesting that the condition of not knowing to which generation one will belong is equivalent to not knowing which stage of "civilization" one will experience).

42. Geoffrey Heal, *Discounting: A Review of the Basic Economics*, 74 U. CHI. L. REV. 59, 60 (2007).

43. *See* Lisa Heinzerling, *Regulatory Costs of Mythic Proportions*, 107 YALE L.J. 1981, 2051 (1998).

44. Louis Kaplow, *Discounting Dollars, Discounting Lives: Intergenerational Distributive Justice and Efficiency*, 74 U. CHI. L. REV. 79, 85 (2007).

45. Cass R. Sunstein & Arden Rowell, *On Discounting Regulatory Benefits: Risk, Money, and Intergenerational Equity*, 74 U. CHI. L. REV. 171, 203 (2007).

46. *See, e.g.,* Arrow et al., *supra* note 27, at 151 ("Even if some resources such as stocks of minerals are drawn down along a consumption path, the sustainability criterion could nevertheless be satisfied if other capital assets were accumulated sufficiently to offset the resource decline."); Robert M. Solow, *An Almost Practical Step toward Sustainability*, 19 RESOURCES POL'Y 162, 168 (1993) ("The duty imposed by sustainability is to bequest to posterity not any particular thing . . . but rather to endow them with whatever it takes to achieve a standard of living at least as good as our own and to look after their next generation similarly.").

47. *See* John Hartwick, *Intergenerational Equity and the Investing of Rents from Exhaustible Resources*, 67 AM. ECON. REV. 972, 973–74 (1977) (noting that "invest[ing] all net returns from exhaustible resources in reproducible capital . . . implies intergenerational equity").

48. For instance, one-time critical legal studies scholar Roberto Unger recently has embraced the technological optimist's view: "We view ourselves as managers, in trust for future generations, of a sinking fund of non-renewable resources. We balance the call of consumption against the duty of thrift. It is an anxiety founded on an illusion. Necessity, mother of invention, has never yet in modern history failed to elicit a scientific and technological response to the scarcity of a resource, leaving us richer than we were before. If the earth itself were to waste away, we would find a way to flee from it into other reaches of the universe." ROBERTO MANGABEIRA UNGER, THE SELF AWAKENED: PRAGMATISM UNBOUND 240 (2007).

49. *See* DALY, *supra* note 40, at 82. Lest one balk at such seemingly draconian measures, consider the fact that even these principles may not be adequate to ensure environmental sustainability: If particular exhaustible resources are sufficiently important and nonsubstitutable within production and consumption, then indefinite sustainability is simply impossible, as students of the laws of thermodynamics already appreciate. *See* PARTHA DASGUPTA & GEOFFREY M. HEAL, ECONOMIC THEORY AND EXHAUSTIBLE RESOURCES 4–5 (1979).

50. WILLARD STERNE RANDALL, THOMAS JEFFERSON: A LIFE 486 (1993) (quoting a letter from Thomas Jefferson to James Madison).

51. Martin L. Weitzman, *Gamma Discounting,* 91 Am. Econ. Rev. 260, 266–69 (2001) (describing a survey of 2,160 economists showing that the majority would use lower discount rates for long-term projects than for short-term projects).

52. Heal, *supra* note 38, at 20.

53. Bjørn Lomborg, The Skeptical Environmentalist: Measuring the Real State of the World 312 (2001).

54. *See* Spash, *supra* note 34, at 157 ("A serious GHG reduction programme would alter the technological base of the economy, *e.g.,* developing alternative energy sources, new transportation systems and lifestyles. The basis for comparison of winners and losers is then no longer identifiable. This affects both benefit estimation and cost analysis.").

55. John M. Hartwick, *supra* note 47, at 973–74.

56. P. K. Rao, Sustainable Development: Economics and Policy 105 (2000).

57. Robert M. Solow, *Sustainability: An Economist's Perspective,* in Economics of the Environment 179, 181 (Robert Dorfman & Nancy S. Dorfman eds., 1993).

58. Dexter Samida & David A. Weisbach, *Paretian Intergenerational Discounting,* 74 U. Chi. L. Rev. 145, 147 (2007).

59. *Id.* at 150, 152.

60. Sunstein & Rowell, *supra* note 45, at 182.

61. Viscusi, *supra* note 8, at 230.

62. Sunstein & Rowell, *supra* note 45, at 175 n.26.

63. Samida & Weisbach, *supra* note 58, at 165.

64. Eric Neumayer, *Global Warming: Discounting Is Not the Issue, but Substitutability Is,* 27 Energy Pol'y 33, 40 (1999).

65. Samida & Weisbach, *supra* note 58, at 153.

66. John Ruskin, "Unto This Last": Four Essays on the First Principles of Political Economy 88 (Lloyd J. Hubenka ed., Univ. of Neb. 1967) (1862).

67. *See* Parfit, *supra* note 24, at 483.

68. Samida & Weisbach, *supra* note 58, at 164–65.

69. *See* John Broome, *Discounting the Future,* 23 Phil. & Pub. Aff. 128, 149 n.17 (1994).

70. Sunstein & Rowell, *supra* note 45, at 183; *see also* Sunstein, *supra* note 41, at 285 (contending that intergenerational equity and efficiency "must be analyzed separately").

71. Revesz, *supra* note 38, at 1008–9.

72. Samida & Weisbach, *supra* note 58, at 151 n.24.

73. Revesz, *supra* note 38, at 1008–9.

74. These quotes come, respectively, from Frank Ramsey, *A Mathematical Theory of Saving,* 38 Econ. J. 543, 543 (1928); Arthur C. Pigou, The Economics of Welfare 25 (1932); R. F. Harrod, Towards a Dynamic Economics 40 (1948); *see also* Henry Sidgwick, The Methods of Ethics 414 (Macmillan,

7th ed. 1907) (noting that "the time at which a man exists cannot affect the value of his happiness from a universal point of view").

75. Samida & Weisbach, *supra* note 58, at 153.

76. Talbot Page, *On the Problem of Achieving Efficiency and Equity, Intergenerationally*, 73 L. AND ECON. 580, 591 (1997); *see also* MARK SAGOFF, THE ECONOMY OF THE EARTH: PHILOSOPHY, LAW, AND THE ENVIRONMENT 63 (1988) (observing that "[o]ur decisions concerning the environment will . . . determine, to a large extent, what future people are like and what their preferences and tastes will be").

77. JACQUES DERRIDA, ADIEU TO EMMANUEL LEVINAS 72 (Pascale-Anne Brault & Michael Naas trans., 1999) (quoting Levinas).

78. *See generally* JACQUES DERRIDA, OF HOSPITALITY (2000).

79. *Cf.* BRIAN BARRY, DEMOCRACY, POWER, AND JUSTICE: ESSAYS IN POLITICAL THEORY 500 (1989) ("It is true that we do not know what the precise tastes of our remote descendants will be, but they are unlikely to include a desire for skin cancer, soil erosion, or the inundation of all low-lying areas as a result of the melting of the ice-caps."); Gregory Kavka, *The Futurity Problem*, in OBLIGATIONS TO FUTURE GENERATIONS 186, 189 (R. I. Sikora & Brian Barry eds., 1978) ("For we do know with high degree of certainty the basic biological and economic needs of future generations—enough food to eat, air to breathe, space to move in, and fuel to run machines.").

CHAPTER 7. OTHER FORMS OF LIFE

1. *See* DEREK PARFIT, REASONS AND PERSONS 351–79 (1984); *see also* Anthony D'Amato, *Do We Owe a Duty to Future Generations to Preserve the Global Environment?* 84 AM. J. INT'L L. 190 (1990); Edith Brown Weiss, *Our Rights and Obligations to Future Generations for the Environment*, 84 AM. J. INT'L L. 198 (1990). For an attempt to resolve the non-identity problem by recourse to Rawls's original position device, see Jeffrey Reiman, *Being Fair to Future People: The Non-Identity Problem in the Original Position*, 35 PHIL. & PUB. AFF. 69 (2007).

2. *See* PARFIT, *supra* note 1, at 370–71.

3. *See* Stephen A. Marglin, *The Social Rate of Discount and the Optimal Return of Investment*, 77 Q. J. ECON. 95, 97 (1963) ("[I] want the government's social welfare function to reflect only the preferences of present individuals. Whatever else democratic theory may or may not imply, I consider it axiomatic that a democratic government reflects only the preferences of the individuals who are presently members of the body politic."); Terence Ball, *The Incoherence of Intergenerational Justice*, 28 INQUIRY 321, 322 (1985) ("We cannot know what men and women in distant generations will mean by 'justice,' nor what they will regard as (un)just. . . . The more distant the generations, the greater the likelihood that their moral concepts and ours will be at least partially or even perhaps wholly incommensurable.").

4. *See* Thomas Schwartz, *Obligations to Posterity,* in OBLIGATIONS TO FUTURE GENERATIONS 3 (R. I. Sikora & Brian Barry eds., 1978).

5. R. I. Sikora, *Is It Wrong to Prevent the Existence of Future Generations? in* OBLI-GATIONS TO FUTURE GENERATIONS, *supra* note 4, at 112, 124 (referring to a common but, in the author's view, unduly simplistic approach); *see also* Jan Narveson, *Future People and Us,* in OBLIGATIONS TO FUTURE GENERATIONS, *supra* note 4, at 38 ("What, if anything, do we owe to future generations? An-swers to this question vary widely. Indeed, they range all the way from Noth-ing to Everything—which would be no cause for alarm, except that both answers, and some in between, have rational support.").

6. *See* PARFIT, *supra* note 1, at 377–78.

7. Eric Posner, *Agencies Should Ignore Distant-Future Generations,* 74 U. CHI. L. REV. 139, 140 (2007).

8. *Id.* at 141 (observing that responsible officials will only "support policies that benefit nonvoting future generations only to the extent that they are sup-ported by voting members of the current generation").

9. *See* MATTHEW D. ADLER & ERIC A. POSNER, NEW FOUNDATIONS OF COST-BENEFIT ANALYSIS 127–35 (2006).

10. Posner, *supra* note 7, at 143 (emphasis added).

11. *Id.* at 142.

12. Seyla Benhabib, *Democracy and Difference: Reflections on the Metapolitics of Lyotard and Derrida,* in THE DERRIDA-HABERMAS READER 128, 136 (Lasse Thomassen ed., 2006).

13. *See* Michael Walzer, *Membership,* in COMMUNITARIANISM AND INDIVIDUALISM 65, 65 (Shlomo Avineri & Avner de-Shalit eds., 1992) ("[W]hat we do with re-gard to membership structures all our other distributive choices: it deter-mines with whom we make those choices, from whom we require obedience and collect taxes, to whom we allocate goods and services.").

14. *See generally* AMARTYA K. SEN, COLLECTIVE CHOICE AND SOCIAL WELFARE (1970); KENNETH J. ARROW, SOCIAL CHOICE AND INDIVIDUAL VALUES (2d ed. 1963).

15. *Cf.* Jacques Derrida, *Declarations of Independence,* 7 NEW POLI. SCI. 7 (1986) (treating a similar co-origination paradox within the Declaration of Indepen-dence). For a collection of insightful essays concerning this theme, see THE PARADOX OF CONSTITUTIONALISM: CONSTITUENT POWER AND CONSTITU-TIONAL FORM (Martin Loughlin & Neil Walker eds., 2007).

16. *Cf.* SEYLA BENHABIB, ANOTHER COSMOPOLITANISM 33 (2006) (observing that "every act of self-legislation is also an act of self-constitution," and that " 'We, the people' who agree to bind ourselves by these laws, are also defining our-selves as a 'we' in the very act of self-legislation").

17. Union of Concerned Scientists, "World Scientists' Warning to Humanity" (1992), *available at* http://www.ucsusa.org/about/1992-world-scientists.html (statement organized by the Union of Concerned Scientists and endorsed by

more than seventeen hundred scientists, including a majority of the living Nobel laureates in the sciences).

18. Judy Clark, Jacquelin Burgess & Carolyn M. Harrison, *"I Struggled with This Money Business": Respondents' Perspectives on Contingent Valuation*, 33 ECOL. ECON. 45, 50 (2000).

19. 42 U.S.C.A. § 4331(a) (2009).

20. 437 U.S. 153, 185 (1978).

21. *Id.* at 184.

22. CHARLES TAYLOR, MODERN SOCIAL IMAGINARIES 17 (2004).

23. For an intellectual history of these movements, see RODERICK FRAZIER NASH, THE RIGHTS OF NATURE: A HISTORY OF ENVIRONMENTAL ETHICS 42 (1989).

24. JEREMY BENTHAM, AN INTRODUCTION TO THE PRINCIPLES OF MORALS AND LEGISLATION 310–11 n.1 (Clarendon Press 1907) (1789) (emphasis in original).

25. *See* John Locke, *Some Thoughts Concerning Education* (1693), in THE EDUCATIONAL WRITINGS OF JOHN LOCKE 225–26 (James L. Axtell ed., 1968); Immanuel Kant, *Duties Towards Animals and Spirits*, in LECTURES ON ETHICS 239, 241 (Louis Infield trans., 1930).

26. Peter Singer, *All Animals are Equal*, in APPLIED ETHICS 215, 225 (Peter Singer ed., 1986).

27. *See, e.g.*, BRIAN BARRY, DEMOCRACY, POWER, AND JUSTICE: ESSAYS IN POLITICAL THEORY 495 (1989) (observing that parity of power is often taken to be a prerequisite for justice relations and that "[a] truistic but fundamental difference between our relations with our successors and our relations with our contemporaries . . . is the absolute difference in power").

28. *See* JOHN RAWLS, A THEORY OF JUSTICE 512 (1971) (noting that "right conduct in regard to animals and the rest of nature" is "outside the scope of the theory of justice, and it does not seem possible to extend the contract doctrine so as to include them in a natural way").

29. *See* J. S. MILL, UTILITARIANISM, LIBERTY AND REPRESENTATIVE GOVERNMENT 208 (J. M. Dent & Sons, 1964) ("The rights and interests of every or any person are only secure from being disregarded when the person interested is himself able, and habitually disposed, to stand up for them.").

30. RICHARD A. POSNER, THE ECONOMICS OF JUSTICE 76 (1983) ("Animals count, but only insofar as they enhance wealth. The optimal population of sheep is determined not by speculation on their capacity for contentment relative to people, but by the intersection of the marginal product and marginal cost of keeping sheep.").

31. *See* Brian Barry, *Circumstances of Justice and Future Generations*, in OBLIGATIONS TO FUTURE GENERATIONS, *supra* note 4, at 204, 205 (observing that "Hume's theory of the circumstances of justice does quite clearly entail that animals and future generations are outside the scope of justice"); DAVID HEYD, GENETHICS: MORAL ISSUES IN THE CREATION OF PEOPLE 50 (1992) ("Since Hobbes and Hume these background conditions making justice a relevant political ideal have been listed as rough equality in natural powers,

vulnerability, interdependence, limited mutual sympathy, and so on. [But people] of different generations cannot cooperate with each other, they cannot have a Humean sympathy for each other, they are not symmetrically vulnerable to harm, and even their resources are not limited in a defined and fixed way.").

32. DAVID HUME, AN ENQUIRY CONCERNING THE PRINCIPLES OF MORALS, AP-PENDIX III, HUME'S MORAL AND POLITICAL PHILOSOPHY 190–91 (H.D. Aiken ed., Hafner 1949) (1751).

33. Singer, *supra* note 26, at 222.

34. *Id.* at 228.

35. BRUNO LATOUR, WE HAVE NEVER BEEN MODERN 24 (Catherine Porter trans., 1993).

36. *Cf.* Kenneth E. Goodpaster, *On Being Morally Considerable,* 75 J. PHIL. 308, 321 (1978) ("Let me hazard the hypothesis, then, that there is a nonaccidental affinity between a person's or a society's conception of value and its conception of moral considerability.").

37. *See, e.g.,* TOM REGAN, DEFENDING ANIMALS RIGHTS (2001).

38. R. G. Frey, *Animals,* in THE OXFORD HANDBOOK OF PRACTICAL ETHICS 161, 174 (Hugh LaFollette ed., 2003).

39. Thomas Scanlon makes this idea plain in connection with an argument regarding the conditions that are necessary to admit nonconventional interest holders into political theory via a trusteeship or guardianship mechanism: "One further minimum requirement for this notion [of trusteeship] is that the being constitute a point of view; that is, that there be such a thing as what it is like to be that being, such a thing as what the world seems like to it. Without this, we do not stand in a relation to the being that makes even hypothetical justification to it appropriate." T. M. Scanlon, *Contractualism and Utilitarianism,* 103, 114, in UTILITARIANISM AND BEYOND 103, 114 (Amartya Sen & Bernard Williams eds., 1982).

40. *See generally* Christine M. Korsgaard, "Fellow Creatures: Kantian Ethics and Our Duties to Animals" (Tanner Lecture on Human Values, University of Michigan, Feb. 6, 2004), *available at* http://www.people.fas.harvard .edu/~ korsgaar/CMK.FellowCreatures.pdf.

41. *But see* FRANCIS FUKUYAMA, OUR POSTHUMAN FUTURE: CONSEQUENCES OF THE BIOTECHNOLOGY REVOLUTION 77 (2002) (outlining reasons to be skeptical that germ-line engineering of the human will happen anytime soon).

42. *See generally* RONALD DWORKIN, LIFE'S DOMINION: AN ARGUMENT ABOUT ABORTION, EUTHANASIA, AND INDIVIDUAL FREEDOM (1994).

43. HEYD, *supra* note 31, at 68 (noting that it is "the way we want the world and future people to be which is the subject of genethical choices").

44. *President Bush's State of the Union Address,* WASH. POST, Jan. 31, 2006 at A1.

45. UNESCO, "Declaration on the Responsibilities of the Present Generations Towards Future Generations," Nov. 12, 1997, UNESCO General Conference, 29th Sess., *available at* http://www.unesco.org/cpp/uk/declarations/genera tions.pdf. *See also* UNESCO, "Universal Declaration on the Human Genome

and Human Rights," Nov. 11, 1997, UNESCO General Conference, 29th Sess., at Article 11 ("Practices which are contrary to human dignity, such as reproductive cloning of human beings, shall not be permitted."), *available at* http://portal.unesco.org/shs/en/ev.php-URL_ID=1881&URL_DO=DO_TOPIC&URL_SECTION=201.html.

46. JÜRGEN HABERMAS, THE FUTURE OF HUMAN NATURE 47 (2003).

47. *Id.* at 25.

48. *Id.* at 82.

49. *Id.* at 65; *see also* BRUCE A. ACKERMAN, SOCIAL JUSTICE IN THE LIBERAL STATE 117 (1980) (stating that "no parent has a right to view 'his' child as a mere instrument for the gratification of his personal conception of the good").

50. HABERMAS, *supra* note 46, at 96; *see also* FUKUYAMA, *supra* note 41, at 208 (also relying on a distinction between therapy and enhancement and expressing confidence that regulators can be relied upon to make the distinction, even conceding the social construction of pathology and disease).

51. DAVID WOOD, THE STEP BACK: ETHICS AND POLITICS AFTER DECONSTRUCTION 4 (2005).

52. *Id.* at 139. At least one philosophically inclined economist has made a similar observation. *See* BRIAN J. LOASBY, KNOWLEDGE, INSTITUTIONS AND EVOLUTION IN ECONOMICS 12 (1999) (observing that absent "responsibility" or other means "to impose closure by non-logical processes," decisions will not be made).

53. SIMON CRITCHLEY, THE ETHICS OF DECONSTRUCTION: DERRIDA AND LEVINAS 5 (2d. ed. 1999) (quoting Levinas).

54. JACQUES DERRIDA, ADIEU TO EMMANUEL LEVINAS 7 (Pascale-Anne Brault & Michael Naas trans., 1999) (quoting Levinas).

55. CRITCHLEY, *supra* note 53, at 48 ("Ethics begins as a relation with a singular, other person who calls me into question and then, and only then, calls me to the universal discourse of reason and justice. Politics begins as ethics.").

56. EMMANUEL LEVINAS, TIME AND THE OTHER 33 (Richard A. Cohen trans., 1987).

57. *Id.*

58. IROQUOIS NATIONS CONSTITUTION, Tree of the Long Leaves § LI, reprinted in ARTHUR C. PARKER, THE CONSTITUTION OF THE FIVE NATIONS 38–39 (Iroqrafts Ltd. 1991) (1916).

59. The passage reads: "Let us suppose that the great empire of China, with all its myriads of inhabitants, was suddenly swallowed up by an earthquake, and let us consider how a man of humanity in Europe . . . would be affected upon receiving intelligence of this dreadful calamity. He would, I imagine, first of all, express very strongly his sorrow for the misfortune of that unhappy people. . . . And when all this fine philosophy was over, when all these humane sentiments had been once fairly expressed, he would pursue his business or his pleasure, take his repose or his diversion with the same ease and tranquility, as if no such accident had happened. . . . If he was to lose his little finger to-morrow, he would not sleep to-night; but . . . he will snore with the most

profound security over the ruin of a hundred millions of his brethren. . . ." ADAM SMITH, THE THEORY OF MORAL SENTIMENTS 136 (bk. III, ch. 3, § 4) (D. D. Raphael & A. L. Macfie eds., 1976)

60. Jacques Derrida, *The Animal That Therefore I Am (More to Follow)*, 28 CRITI-CAL INQUIRY 369, 392 (David Wills trans., 2002) ("The animal is a word, it is an appellation that men have instituted, a name they have given themselves the right and the authority to give to another living creature."). Although Levinas's writings are amenable to a more-inclusive reading, most commentators take his ethics to remain restricted to humans, especially given later work in which the philosopher emphasized the contrast between a subject's inassimilable capacity of authorship and the determinacy of the subject's vocalizations—between, in Levinas's terms, the "Saying" and the "Said." *See, e.g.*, MATTHEW CALARCO, ZOOGRAPHIES: THE QUESTION OF THE ANIMAL FROM HEIDEGGER TO DERRIDA 55 (2008). Indeed, Derrida called it "a matter for serious concern" that in Levinas's construction of the face-to-face relation, "he is speaking of man." Derrida, *supra* note 60, at 381.

61. *Id.* at 379.

62. *Id.* at 382.

63. *Id.* at 387.

64. CALARCO, *supra* note 60, at 71.

65. *See generally* K. E. LØGSTRUP, BEYOND THE ETHICAL DEMAND (2007).

66. WOOD, *supra* note 51, at 68 (emphasis in original removed). The degree to which Levinas himself is guilty of this excessive abstraction is a matter of dispute. On Simon Critchley's reading, for instance, "[e]thics is not an obligation towards the Other mediated through the formal and procedural universalization of maxims or some appeal to good conscience; rather—and this is what is truly provocative about Levinas—ethics is lived as a corporeal obligation to the Other, an obligation whose form is sensibility." CRITCHLEY, *supra* note 53, at 180. To Wood, however, the Levinasian refusal to adopt any ontological shortcuts whatsoever in the face of the unknowability of the other renders the project too wispy to guide lived experience. *See also* JÜRGEN HABERMAS, BETWEEN NATURALISM AND RELIGION 280 (Ciaran Cronin trans., 2008) (expressing doubt that the Levinasian understanding of the face-to-face relation can ground legal obligations).

67. LEVINAS, *supra* note 56, at 42.

CHAPTER 8. ECOLOGICAL RATIONALITY

1. EMMANUEL LEVINAS, TOTALITY AND INFINITY: AN ESSAY ON EXTERIORITY 199 (1969).

2. *See* ALFRED NORTH WHITEHEAD, SCIENCE AND THE MODERN WORLD 201 (1925) (1997) (stating that "the true rationalism must always transcend itself by recurrence to the concrete in search of inspiration").

3. Final Regulations to Establish Requirements for Cooling Water Intake Structures at Phase II Existing Facilities, 69 Fed. Reg. 41,576, 41,586 (July 9, 2004).

4. *Id.*

5. 33 U.S.C.A. 1326(b) (2009).

6. *See Appalachian Power Co. v. Train*, 566 F.2d 451, 457 (4th Cir. 1977); National Pollutant Discharge Elimination System; Revision of Regulations, 44 Fed. Reg. 32,956 (June 7, 1979).

7. *See Cronin v. Browner*, 898 F. Supp. 1052, 1056 (S.D.N.Y. 1995).

8. *See* Karl R. Rábago, *What Comes Out Must Go In: Cooling Water Intakes and the Clean Water Act*, 16 HARV. ENVTL. L. REV. 429, 467 (1992).

9. *See Riverkeeper, Inc. v. U.S. EPA*, 358 F.3d 174, 181 (2d Cir. 2004) (*Riverkeeper I*); *Riverkeeper, Inc. v. U.S. EPA*, 475 F.3d 83, 90 (2d Cir. 2007) (*Riverkeeper II*).

10. 69 Fed. Reg. 41,576, 41,598.

11. *Id.* at 41,601. *See also* United States Environmental Protection Agency, Economic and Benefits Analysis for the Final Section 316(b) Phase II Existing Facilities Rule, EPA-821-R-04-005, Feb. 2004, at A2-5 ("Closed-cycle cooling systems . . . are the most effective means of protecting organisms from [impingement and entrainment]."), *available at* http://www.epa.gov/waterscience/316b/phase2/econbenefits/final.htm.

12. 69 Fed Reg. 41,576, 41,583.

13. *See id.* at 41,603 ("[I]n determining that the technologies on which EPA based the compliance alternatives and performance standards are the best technologies available for existing facilities to minimize adverse environmental impact, EPA considered the national cost of those technologies in comparison to the national benefits. . . .").

14. United States Environmental Protection Agency, *supra* note 11, at D1-1 (predicting "no compliance action" for two hundred facilities but noting that seventy-five of these facilities already have adopted closed-cycle cooling).

15. *Id.* at C3-2.

16. As Richard Stewart has noted, one frequently offered interpretation of the precautionary principle is that the best-available-pollution-control technology should be required of all proponents of activities with uncertain environmental, health, or safety threats. *See* Richard B. Stewart, *Environmental Regulatory Decision Making Under Uncertainty*, in 20 RES. IN L. AND ECON. 71, 78 (Timothy Swanson ed., 2002); *see also* Daniel Bodansky, *The Precautionary Principle in U.S. Environmental Law*, in INTERPRETING THE PRECAUTIONARY PRINCIPLE (Timothy O'Riordan & James Cameron eds., 1994); Adam Babich, *Too Much Science in Environmental Law*, 28 COLUM. J. ENVTL. L. 119, 125 (2003) ("The requirement of best available technology embodies a policy judgment as attractive as apple pie."); Howard Latin, *Ideal Versus Real Regulatory Efficiency: Implementation of Uniform Standards and "Fine-Tuning" Regulatory Reforms*, 37 STAN. L. REV. 1267, 1283–84 (1985) (noting Congress repeatedly "chose to emphasize the need for prompt injury prevention over the need for an optimal balance between regulatory benefits and costs" in its landmark 1970s' legislation); Thomas O. McGarity, *The Goals of Environmental Legislation*, 31

B.C. ENVTL. AFF. L. REV. 529, 538–45 (2004) (reviewing examples); Wendy E. Wagner, *The Triumph of Technology-Based Standards*, 2000 U. ILL. L. REV. 83 (attributing much of the success of pollution reduction in the modern environmental era to technology-based standards). Significant early environmental court decisions also emphasized the precautionary basis of U.S. risk regulation. *See Reserve Mining Co. v. E.P.A.*, 514 F.2d 492 (8th Cir. 1975); *Ethyl Corp. v. E.P.A.*, 541 F.2d 1, 28 (D.C. Cir. 1976).

17. *Cf.* Gerd Gigerenzer, *Heuristics*, in HEURISTICS AND THE LAW 17, 23 (Gerd Gigerenzer & Christoph Engel eds., 2006) ("[T]he rationality of a heuristic is external or 'ecological' (i.e., how well a heuristic performs in a real-world environment) and not internal.").

18. Robert N. Stavins, Letter to Proposed Rule Comment Clerk—W-00-32, Re: Comments on Proposed Rule RIN 2040-AD62 Clean Water Act Section 316(b)—National Pollutant Discharge Elimination System—Proposed Regulations for Cooling Water Intake Structures at Phase II Existing Facilities, EPA ICR no. 2060.01, at 9 (July 19, 2002).

19. *Id.* at 3.

20. *Id.* at 31, 42.

21. *Id.* at 34.

22. *See* 69 Fed. Reg. 41,603 ("Section 316(b) authorizes consideration of the environmental benefit to be gained by requiring that the location, design, construction, and capacity of cooling water intake structures reflect the best economically *practicable* technology available for the purpose of minimizing adverse environmental impact.") (emphasis added).

23. *See Entergy Corp. v. Riverkeeper, Inc.*, 129 S.Ct. 1498 (2009).

24. Stavins, *supra* note 18, at 8.

25. *Id.* at 9.

26. *Id.*

27. 69 Fed. Reg. 41,576, 41,586.

28. United States Environmental Protection Agency, Regional Analysis Document for the Final Section 316(b) Phase II Existing Facilities Rule, EPA-821-R-02-003, Feb. 12, 2004, at A9-1, *available at* http://www.epa.gov/waterscience/316b/phase2/casestudy/final.htm.

29. United States Environmental Protection Agency, *supra* note 11, at C3-2; *see also* 69 Fed. Reg. 41,661 ("The Agency's direct use valuation does not account for the benefits from the remaining 98.2% of the age 1 equivalent aquatic organisms estimated to be protected nationally under today's rule.").

30. United States Environmental Protection Agency, Case Study Analysis for the Proposed Section 316(b) Phase II Existing Facilities Rule, EPA-821-R-02-001, Feb. 2002, at A8-1 (emphasis added), *available at* http://www.epa.gov/water science/316b/phase2/casestudy/.

31. United States Environmental Protection Agency, *supra* note 28, at A9-8.

32. United States Environmental Protection Agency, *supra* note 11, at D1-5.

33. 69 Fed. Reg. 41,576, 41,660 (quoting Myrick Freeman).

34. United States Environmental Protection Agency, Economic and Benefits Analysis for the Proposed Section 316(b) Phase II Existing Facilities Rule, EPA-821-R-02-001, Feb. 2002, at D1-4, *available at* http://www.epa.gov/water-science/316b/phase2/econbenefits/.

35. Stavins, *supra* note 18, at 10 (emphasis added).

36. Robert N. Stavins, Letter to Water Docket ID No. OW-2002-0049, Re: Comments on Proposed Rule RIN 2040-AD62; Notice of Data Availability Clean Water Act Section 316(b)—National Pollutant Discharge Elimination System—Proposed Regulations to Establish Requirements for Cooling Water Intake Structures at Phase II Existing Facilities, at 4 (Apr. 21, 2003); *see also id.* at 2 (claiming that Ackerman offered views that "intermix and thereby confuse [methodological] issues with personal value judgments about the benefits of environmental resources").

37. *See, e.g.,* James Cameron, *The Precautionary Principle in International Law,* in REINTERPRETING THE PRECAUTIONARY PRINCIPLE 113, 119 (Tim O'Riordan, James Cameron & Andrew Jordan eds., 2001) ("[B]y explicitly noting the limits of scientific determination, the precautionary principle legitimates public political determination of these issues, in some sense democratising international environmental law and its implementation.").

38. Frank Ackerman, Comments on Proposed Rule, RIN 2040-AD62 Clean Water Act section 316(b)—"National Pollutant Discharge Elimination System—Proposed Regulations for Cooling Water Intake Structures at Phase II Existing Facilities," EPA ICR no. 2060.01, Aug. 1, 2002, at 11–13.

39. Frank Ackerman and Rachel Massey, Comments on Notice of Data Availability, EPA 40 CFR Part 125 Clean Water Act section 316(b)—"National Pollutant Discharge Elimination System—Proposed Regulations for Cooling Water Intake Structures at Phase II Existing Facilities," Notice of Data Availability, Mar. 19, 2003, June 2, 2003, at 4.

40. Stavins, *supra* note 36, at 3 (emphasis added).

41. *Id.* at 2.

42. A similar confusion surrounded the issue of whether to assume a level of entrainment survival for fish drawn into cooling water systems. Although some evidence had indicated that not all entrained fish perish, the EPA concluded that the evidence was too uncertain and fragmentary to support any specific survival ratio; thus, the agency assumed a 0 percent survival rate, an assumption that attracted vigorous critiques from industry commentators. As the Second Circuit noted, however, " '[i]t is within EPA's discretion to decide that in the wake of uncertainty, it would be better to give the values a conservative bent rather than err on the other side.' " *Riverkeeper II,* 475 F.3d at 127 (quoting *Am. Iron & Steel Inst. v. EPA,* 115 F.3d 979, 993 (D.C. Cir. 1997)).

43. 69 Fed. Reg. 41,576, 41,624.

44. *See* Amy Sinden, *In Defense of Absolutes: Combating the Politics of Power in Environmental Law,* 90 IOWA L. REV. 1405 (2005).

45. 69 Fed. Reg. 41,576, 41,586.

46. *Id.* at 41,624.

47. *Id.* at 41,587 (observing that threatened, endangered, and other special status species might be impacted by cooling water intake and noting, as an example, "that 3,200 threatened or endangered sea turtles entered enclosed cooling water intake canals at the St. Lucie Nuclear Generating Plant in Florida").

48. *Id.* at 41,612.

49. Stavins, *supra* note 18, at 3, 15.

50. *Id.*

51. 33 U.S.C.A. § 1251(a); *see also Rapanos v. United States* 547 U.S. 715 (2006).

52. Stavins, *supra* note 18, at 17.

53. 69 Fed. Reg. 41,576, 41,610.

54. *Id.*

55. National Pollutant Discharge Elimination System—Proposed Regulations to Establish Requirements for Cooling Water Intake Structures at Phase II Existing Facilities, 67 Fed. Reg. 17,122, 17,138 (proposed Apr. 9, 2002). For instance, one study using conservative entrainment survival assumptions estimated that each year up to 20 percent of striped bass, 25 percent of bay anchovy, and 43 percent of Atlantic tomcod would be lost to entrainment in the Hudson River estuary from just three power plants. *See id.*

56. United States Environmental Protection Agency, *supra* note 28, at A13-12.

57. 69 Fed. Reg. 41,576, 41,624–25.

58. Stavins, *supra* note 18, at 3.

59. *Id.* at 27.

60. *Id.*

61. Stavins, *supra* note 36, at 16 n.36.

62. *See* MARK SAGOFF, THE ECONOMY OF THE EARTH: PHILOSOPHY, LAW, AND THE ENVIRONMENT 7–14 (1988).

63. See Stavins, *supra* note 36, at 3 (contending that "the cost of a good or service cannot be used as a proxy for its respective benefits except in those situations where individuals (*or groups*) have been observed to voluntarily purchase the good or service") (emphasis added).

64. Stavins, *supra* note 18, at 25 (emphasis in original).

65. See Executive Order No. 13,422, 72 Fed. Reg. 2763, 2763–65 (Jan. 18, 2007); Final Bulletin for Agency Good Guidance Practices, 72 Fed. Reg. 3432 (Jan. 25, 2007).

66. Amartya Sen, *The Discipline of Cost-Benefit Analysis,* 29 J. LEGAL STUD. 931, 949 (2000).

67. Amartya Sen, *Why We Should Preserve the Spotted Owl,* LONDON REV. BOOKS, Feb. 5, 2004, *available at* http://www.lrb.co.uk/v26/n03/sen_01_.html.

68. *Id.*

69. *See, e.g., Riverkeeper II,* 475 F.3d at 105, 113 n.25.

70. Stavins, *supra* note 18, at 42.

71. *Id.* at 37.

72. Stavins, *supra* note 36, at 3.

73. *Id.* at 4.

74. *Id.*

75. *Id.* at 4 n.6.

76. Such claims by welfare economists abound. Another memorable example came from William Nordhaus, who argued that the use of a laissez-faire base-line for the quantification of climate change policy impacts was an observational posture akin to the natural sciences: "[T]he baseline model is an attempt to project from a positive perspective the levels and growth of population, output, consumption, saving, interest rates, greenhouse gas emissions, climate change, and climate damages as would occur with no interventions to affect greenhouse gas emissions. This approach does not make a case for the social desirability of the distribution of incomes over space or time of existing conditions, any more than a marine biologist makes a moral judgment on the equity of the eating habits of marine organisms in attempting to understand the effect of acidification on marine life." William D. Nordhaus, *A Review of* The Stern Review on the Economics of Climate Change, 45 J. ECON. LIT. 686, 692 (2007).

77. Stavins, *supra* note 36, at 5.

78. Stavins, *supra* note 18, at 20 ("There is a consensus view in economics that when an appropriate and reliable revealed-preference approach is available for valuing a particular environmental amenity, then that approach should be used, rather than resorting to a stated preference approach, such as contingent valuation.").

79. *Id.* at 17.

CHAPTER 9. ENVIRONMENTAL CONSTITUTIONALISM

1. In the majority of the world's constitutions the environment is given express constitutional significance through various formulations of a right to a healthy and sustainable environment or through a governmental obligation to protect the environment. Although exact accounting differs by commentator, according to Tim Hayward, "around fifty" nations' constitutions contain environmental rights, and "more than a hundred countries have constitutional environmental provisions of some kind." TIM HAYWARD, CONSTITUTIONAL ENVIRONMENTAL RIGHTS 3–4 (2005).

2. *See* Robert V. Percival, *"Greening" the Constitution—Harmonizing Environmental and Constitutional Values*, 32 ENVTL. L. 809 (2002).

3. *See* Barton H. Thompson Jr., *Constitutionalizing the Environment: The History and Future of Montana's Environmental Provisions*, 64 MONT. L. REV. 157, 158 (2003) (observing that "[m]ore than a third of all [U.S.] state constitutions, including all written since 1959, address modern concerns of pollution and resource preservation").

4. *See* Barton H. Thompson Jr., *Environmental Policy and State Constitutions: The Potential Role of Substantive Guidance*, 27 RUTGERS L.J. 863, 897 (1996) (noting that "[c]ourts actively have sought out legal justifications for avoiding en-

vironmental policymaking even in those states with strong environmental policy provisions that appear on their face to impose mandates or obligations on the legislature or directly on the regulated community").

5. For a compilation of sources, see Douglas A. Kysar, *The Consultants' Republic,* 121 HARV. L. REV. 2041, 2060–62 (2008).

6. For examples of this earlier scholarship, see, respectively, Christopher D. Stone, *Should Trees Have Standing?—Toward Legal Rights for Natural Objects,* 45 S. CAL. L. REV. 450 (1972); Joseph L. Sax, *The Public Trust Doctrine in Natural Resource Law: Effective Judicial Intervention,* 68 MICH. L. REV. 471 (1970); EDITH BROWN WEISS, IN FAIRNESS TO FUTURE GENERATIONS: INTERNATIONAL LAW, COMMON PATRIMONY, AND INTERGENERATIONAL EQUITY (1989); Luis Kutner, *The Control and Prevention of Transnational Pollution: A Case for World Habeas Ecologicus,* 9 LAW. AM. 257 (1977).

7. *Cf.* David A. Westbrook, *Liberal Environmental Jurisprudence,* 27 U.C. DAVIS L. REV. 619, 711 (1994) ("To date, contemporary liberal ideology has tried to appropriate the essentially religious implications of the concept of nature as either personal preference, and hence of highly limited importance for politics, or as objective truth, certified by the new science, and hence profoundly alienated from individual experience.").

8. *See* Carol M. Rose, *Environmental Law Grows Up (More or Less), and What Science Can Do to Help,* 9 LEWIS & CLARK L. REV. 273, 293 (2005) (advocating continued reliance on science as the fundamental basis of environmental law); A. Dan Tarlock, *Environmental Law: Ethics or Science?* 7 DUKE ENVTL. L. & POL'Y FORUM 193, 223 (1996) (concluding that environmental law must remain science based); Paul Wapner, *Environmental Ethics and Global Governance: Engaging the International Liberal Tradition,* 3 GLOBAL GOVERNANCE 213, 213 (1997) (suggesting the reframing of international environmental concerns in the nationalist and individualist terms liberalism cognizes).

9. Thompson, *supra* note 3, at 187.

10. *Id.* at 198; *see also* A. Dan Tarlock, *Environmental Law, but Not Environmental Protection, in* NATURAL RESOURCES POLICY AND LAW: TRENDS AND DIRECTIONS 162, 171 (L. J. MacDonnell & S. F. Bates eds., 1993) ("[E]nvironmental law has failed to develop a substantive theory of environmental quality entitlement. The western tradition of expanding the concept of human dignity left no room for the protection of non-human values."); A. Dan Tarlock, *Is There a There There in Environmental Law?* 19 J. LAND USE & ENVTL. L. 213, 223 (2004) (noting that "there is no longstanding social consensus about the central question of modern environmentalism—the 'correct' human stewardship relationship to the natural world") (footnote omitted).

11. *See* THE CONSTITUTION OF THE KINGDOM OF BHUTAN art. 5(3), *available at* http://www.constitution.bt/TsaThrim%20Eng%20(A5).pdf.

12. *See* Robyn Stein, *Water Law in a Democratic South Africa: A Country Case Study Examining the Introduction of a Public Rights System,* 83 TEX. L. REV. 2167 (2005).

13. Barry E. Hill, Steve Wolfson & Nicholas Targ, *Human Rights and the Environment: A Synopsis and Some Predictions*, 16 Geo. Int'l Envtl. L. Rev. 359, 361 (2004). A similar view was expressed by the Second Circuit Court of Appeals in an opinion dismissing foreign plaintiffs' claims under the Alien Tort Claims Act against an alleged polluting copper-mine facility: "[A]s a practical matter, it is impossible for courts to discern or apply in any rigorous, systematic, or legal manner international pronouncements that promote amorphous, general principles." *Flores v. Southern Peru Copper Corp.*, 343 F.3d 140, 158 (2d Cir. 2002).

14. Amy Sinden similarly argues the absolutism of 1970s-era environmental statutes is best understood as an effort to redress pervasive political imbalances in environmental law and policy-making settings. *See* Amy Sinden, *In Defense of Absolutes: Combating the Politics of Power in Environmental Law*, 90 Iowa L. Rev. 1405 (2005).

15. Sinden describes the supermandate as follows: "Part of the Republican Party's 'Contract with America,' the Risk Assessment and Cost-Benefit Act of 1995 would have . . . require[ed] agencies to perform formal cost-benefit analyses of all proposed major 'health, safety, and environmental' rules (those with annual costs of $25 million or more), and prohibited promulgation of any final rule unless the agency 'certified' that the benefits justify the costs. The bill explicitly stated that its provisions were to 'supplement and, to the extent there is a conflict, supersede' the decision criteria for rulemaking otherwise applicable under the statute pursuant to which the rule is promulgated. After it passed the House by 277 to 141, the Senate counterpart fell just two votes short of overcoming a filibuster." *Id.* at 1422 n.54 (2005) (internal citations omitted).

16. *Cf.* David Wood, The Step Back: Ethics and Politics after Deconstruction 185 (2005) ("Every species that dies out is the loss of an adventure with the future. And with such loss of differentiation we also lose ecological complexity, and hence the diminution of constitutional relationality in nature.").

17. Thompson, *supra* note 3, at 166.

18. 988 P.2d 1236 (1999).

19. *Id.* at 1249 ("The delegates did not intend to merely prohibit the degree of environmental degradation which can be conclusively linked to ill health or physical endangerment. Our constitution does not require that dead fish float on the surface of our state's rivers and streams before its farsighted environmental protections can be invoked.").

20. *See Cape-France Enterprises v. Estate of Peed*, 29 P.3d 1011 (2001).

21. *Sunburst School District No. 2 v. Texaco, Inc.*, 2004 WL 5314057 (Mont. Dist.).

22. *Id.* The Montana Supreme Court upheld this ruling on common-law grounds without opining on the degree to which the environmental constitutional provision affords a basis for recovery when common-law remedies are inadequate. *See Sunburst School District No. 2 v. Texaco, Inc.*, 165 P.3d 1079, 1098 (Mont. 2007). As part of its reasoning on the common-law damages ques-

tion, however, the court did observe that "[a] strict cap on restoration damages . . . would fail to provide an adequate remedy for an injury to the environment [given that a]reas with great ecological value may have little or no commercial value." *Id.* at 1090.

23. In a series of important contributions, Catherine O'Neill has tracked and analyzed the EPA's mercury rulemaking efforts with special attention to the agency's flawed approach to environmental justice aspects of the rulemaking. *See* Catherine A. O'Neill, *The Mathematics of Mercury,* in ALTERNATIVE APPROACHES TO REGULATORY IMPACT ANALYSIS: A DIALOGUE BETWEEN ADVOCATES AND SKEPTICS OF COST-BENEFIT ANALYSIS (Winston Harrington, Lisa Heinzerling & Richard Morgenstern eds., forthcoming); Catherine A. O'Neill, *Protecting the Tribal Harvest: The Right to Catch and Consume Fish,* 22 J. ENVTL. L. & LITIGATION 131 (2007); Catherine A. O'Neill, *Mercury, Risk, and Justice,* 34 ENVTL. L. REP. 11,070 (2004).

24. *See North Carolina v. EPA,* 531 F.3d 896 (D.C. Cir. 2008); *New Jersey v. EPA,* 517 F.3d 574 (D.C. Cir. 2008).

25. *See* Catherine A. O'Neill, *Environmental Justice in the Tribal Context: A Madness to EPA's Method,* 38 ENVTL. L. 495, 524 (2008).

26. *See, e.g.,* JOHN STUART MILL, UTILITARIANISM 10 (George Sher ed., Hackett Publishing 1979) (1861) ("It is better to be a human being dissatisfied than a pig satisfied; better to be Socrates dissatisfied than a fool satisfied.").

27. *See generally* Catherine A. O'Neill, *No Mud Pies: Risk Avoidance as Risk Regulation,* 31 VT. L. REV. 273 (2007); Catherine A. O'Neill, *The Perils of Risk Avoidance,* 20 NAT. RES. & ENV'T 9 (Winter 2006); Catherine A. O'Neill, *Risk Avoidance, Cultural Discrimination, and Environmental Justice for Indigenous Peoples,* 30 ECOLOGY L.Q. 1 (2003).

28. *See* Ronald H. Coase, *The Problem of Social Cost,* 3 J. L. & ECON. 1 (1960).

29. *See* Standards for Performance for New and Existing Sources: Electric Utility Steam Generating Units, 70 Fed. Reg. 28,606, 28,648 (May 18, 2005) (to be codified at 40 C.F.R. pts. 60, 72, 75).

30. O'Neill, *supra* note 25, at 520.

31. *See* KENJI YOSHINO, COVERING: THE HIDDEN ASSAULT ON OUR CIVIL RIGHTS(2006).

32. Bernard Williams, *Persons, Character, and Morality,* in MORAL LUCK: PHILOSOPHICAL PAPERS 18 (1981).

33. W. Kip Viscusi, *Risk Equity,* 29 J. LEGAL STUD. 843, 844 (2000).

34. Cass R. Sunstein and Arden Rowell, *On Discounting Regulatory Benefits: Risk, Money, and Intergenerational Equity,* 74 U. CHI. L. REV. 171, 178 (2007).

35. *Cf.* RONALD DWORKIN, LAW'S EMPIRE 189 (1986) (contending that citizens must understand themselves as authors of law).

36. MATTHEW ADLER & ERIC A. POSNER, NEW FOUNDATIONS FOR COST-BENEFIT ANALYSIS 25 (2005).

37. *Id.* at 53.

38. *Id.* at 102.

39. *Id.* at 157. More recently, Posner has gone further to make the startling claim that international human rights law should be redirected toward the promotion of social welfare, despite the obvious and long-standing connection between human rights and the Kantian intellectual tradition. *See* Eric A. Posner, "Human Welfare, Not Human Rights" (Univ. of Chicago Law Sch. John M. Olin Program in Law & Econ., Working Paper No. 394, 2d series, 2008), *available at* http://papers.ssrn.com/sol3/papers.cfm?abstract_id=1105209.

40. *See Whitman v. American Trucking Associations,* 531 U.S. 457 (2001) (rejecting industry's argument that the EPA should consider costs when setting air quality standards under the CAA).

41. 16 U.S.C.A. § 1851(a)(5) (2009).

42. *Cf.* Holly Doremus, *Constitutive Law and Environmental Policy,* 22 STAN. EN-VTL. L.J. 295 (2003) (advocating a "constitutive approach" to environmentalism that calls for explicitly clarifying and debating the goals of environmental law).

43. *See* HERMAN E. DALY, BEYOND GROWTH 82(1996). As Buzz Thompson notes, the Alaska State Constitution purports to require use of the "sustainable yield principle" as a management criterion for all replenishable public resources, although courts have not lent much practical strength to the provision. *See* Thompson, *supra* note 4, at 901–2. It bears noting, as Thompson also points out, that sustainability is a far less precise concept than often portrayed: "Resources can be 'sustainable,' for example, at various stock levels. A sustainability standard thus leaves open to question whether a constitution intends to sustain a high, medium, or low level of a particular resource. Sustainability also leaves open to debate whether it requires preservation only of yield levels or, more logically, also of quality. Even the concept of a 'renewable resource' is open to interpretation, depending on the time frame involved. Virtually all resources, including petroleum, are renewable given a long enough time frame." *Id.* Thompson concludes from this discussion that courts would likely defer to legislative determinations of the content of a constitutional sustainability obligation. Vagueness alone seems unable to drive this conclusion, however, given the equally amorphous nature of due process, equal protection, interstate commerce, and a host of other hallmark constitutional concepts.

44. *Cf.* Joseph L. Sax, *The Unfinished Agenda of Environmental Law,* 14 HASTINGS W.-NW. J. ENVTL. L. & POL'Y 1 (2008) ("It is a sobering thought that while virtually every other interest that we consider vital has been made the subject of enforceable legal rights, our heritage of biodiversity stands largely outside the framework of established jurisprudential theory.").

45. John O'Neill, *Representing People, Representing Nature, Representing the World,* 19 ENV'T & PLANNING C: GOV'T & POL'Y 483, 497 (2001) ("Given the necessary absence of authorisation, accountability, and presence, claims to speak on behalf of nonhumans and future generations relies on epistemic claims, coupled with care.").

46. *Tarlock,* supra *note 10,* at 224.

47. Cass R. Sunstein, Worst-Case Scenarios 65 (2007).

48. *Cf.* Paul Slovic, *"If I Look at the Mass I Will Never Act": Psychic Numbing and Genocide,* 2 Judgment & Decision Making 79 (2007) (describing psychological studies in which both positive affective sentiment and willingness to make donations to humanitarian causes were shown to decline when analytical thinking was primed through statistical information and through increasing the number of individuals benefitted).

49. Bruce Ackerman, *The Living Constitution* (2006 Oliver Wendell Holmes Lectures) 120 Harv. L. Rev. 1737, 1793 (2007).

50. Thompson, *supra* note 4, at 898.

51. *See* Bruce A. Ackerman, Social Justice in the Liberal State 93–95 (1980).

52. *Cf.* Jürgen Habermas, Between Naturalism and Religion 273 (Ciaran Cronin trans., 2008) (describing Rousseauean tradition in which "sovereignty branches internally into a communitarian understanding of the political freedom of the members of a national community and toward the outside into a collectivist understanding of the freedom of a nation that asserts its existence against other nations").

53. Samuel Beckett, Worstward Ho 7 (1983).

54. Jacques Derrida, *Afterword: Toward an Ethic of Discussion,* in Limited Inc, 116 (Samuel Weber & Jeffrey Mehlman trans., Gerald Graff ed., 1988).

55. J. B. Ruhl, *The Metrics of Constitutional Amendments: And Why Proposed Environmental Quality Amendments Don't Measure Up,* 74 Notre Dame L. Rev. 245, 253 (1999).

56. *Id.*

57. *See* Daly, *supra* note 43, at 82.

58. *See generally* Bruce Ackerman, We the People: Foundations (1991); Bruce Ackerman, We the People: Transformations (1998); Ackerman, *supra* note 49. Ackerman is not alone in encouraging scholars to focus on the diverse mechanisms by which constitutional change occurs. For a helpful overview of numerous contributions, see Heather K. Gerken, *The Hydraulics of Constitutional Reform: A Skeptical Response to Our Undemocratic Constitution,* 55 Drake L. Rev. 925 (2007).

59. Ackerman, *supra* note 49, at 1781–82.

60. *Id.* at 1783–84.

61. In part for this reason, scholars often cite 1970s-era federal environmental law as an example of a civic republican moment in which "the elusive voice of the public good, momentarily audible above the din of power politics, carries the day." Sinden, *supra* note 14, at 1447 (citing Jerry L. Mashaw, Greed, Chaos, and Governance: Using Public Choice to Improve Public Law 33 (1997); *see also* Daniel A. Farber, *Politics and Procedure in Environmental Law,* 8 J. L. Econ. & Org. 59, 65–67 (1992); Steven Kelman, *"Public Choice" and Public Spirit,* 87 Public Interest 80, 91 (1987); *but see* Christopher H.

Schroeder, *Rational Choice versus Republican Moment—Explanations for Environmental Laws, 1969–73,* 9 DUKE ENVTL. L. & POL'Y FORUM 29 (1998) (arguing that rational choice/interest group theory can explain the emergence of modern federal environmental law).

62. *See* 42 U.S.C. A.§ 4331(c) (2009) ("The Congress recognizes that each person should enjoy a healthful environment and that each person has a responsibility to contribute to the preservation and enhancement of the environment.").

63. *See* Environmental Protection Agency, "Advance Notice of Proposed Rulemaking: Regulating Greenhouse Gas Emissions under the Clean Air Act," EPA-HQ-OAR-2008-0318, July 11, 2008, *available at* http://www.epa.gov/climatechange/anpr.html.

64. Ackerman, *supra* note 49, at 1805.

65. GIORGIO AGAMBEN, THE OPEN: MAN AND ANIMAL 76–77 (Kevin Attell trans., 2004).

66. *Id.*

67. *Id.*

68. *See* Diane Brady & Christopher Palmeri, *The Pet Economy,* BUSINESS WEEK, Aug. 6, 2007, *available at* http://www.businessweek.com/magazine/content/07_32/b4045001.htm?chan=search.

69. *See* James Vlahos, *Pill-Popping Pets,* N. Y. TIMES MAGAZINE, July 13, 2008, *available at* http://www.nytimes.com/2008/07/13/magazine/13pets-t.html?pagewanted=all.

70. BRUNO LATOUR, WE HAVE NEVER BEEN MODERN (Catherine Porter trans., 1993) (1991).

71. Václav Klaus, *What Is at Risk Is Not the Climate but Freedom,* FINANCIAL TIMES, June 14, 2007, at 9.

72. THOMAS NAGEL, MORTAL QUESTIONS 211 (1979).